AF862962

Engineering Materials

This series provides topical information on innovative, structural and functional materials and composites with applications in optical, electrical, mechanical, civil, aeronautical, medical, bio- and nano-engineering. The individual volumes are complete, comprehensive monographs covering the structure, properties, manufacturing process and applications of these materials. This multidisciplinary series is devoted to professionals, students and all those interested in the latest developments in the Materials Science field, that look for a carefully selected collection of high quality review articles on their respective field of expertise.

Indexed at Compendex (2021) and Scopus (2022)

Ramesh T. Subramaniam · Ramesh Kasi ·
Shahid Bashir · Sachin Sharma Ashok Kumar
Editors

Graphene

Fabrication, Properties and Applications

Springer

Editors
Ramesh T. Subramaniam
Department of Physics, Faculty of Science
Universiti Malaya
Centre for Ionics University of Malaya
Kuala Lumpur, Malaysia

Ramesh Kasi
Department of Physics, Faculty of Science
Universiti Malaya
Centre for Ionics University of Malaya
Kuala Lumpur, Malaysia

Shahid Bashir
Higher Institution Centre of Excellence
(HICoE), UM Power Energy Dedicated
Advanced Centre (UMPEDAC)
Universiti Malaya
Kuala Lumpur, Malaysia

Sachin Sharma Ashok Kumar
Department of Physics, Faculty of Science
Universiti Malaya
Centre for Ionics University of Malaya
Kuala Lumpur, Malaysia

ISSN 1612-1317 ISSN 1868-1212 (electronic)
Engineering Materials
ISBN 978-981-99-1205-6 ISBN 978-981-99-1206-3 (eBook)
https://doi.org/10.1007/978-981-99-1206-3

© The Editor(s) (if applicable) and The Author(s), under exclusive license to Springer Nature Singapore Pte Ltd. 2023
This work is subject to copyright. All rights are solely and exclusively licensed by the Publisher, whether the whole or part of the material is concerned, specifically the rights of translation, reprinting, reuse of illustrations, recitation, broadcasting, reproduction on microfilms or in any other physical way, and transmission or information storage and retrieval, electronic adaptation, computer software, or by similar or dissimilar methodology now known or hereafter developed.
The use of general descriptive names, registered names, trademarks, service marks, etc. in this publication does not imply, even in the absence of a specific statement, that such names are exempt from the relevant protective laws and regulations and therefore free for general use.
The publisher, the authors, and the editors are safe to assume that the advice and information in this book are believed to be true and accurate at the date of publication. Neither the publisher nor the authors or the editors give a warranty, expressed or implied, with respect to the material contained herein or for any errors or omissions that may have been made. The publisher remains neutral with regard to jurisdictional claims in published maps and institutional affiliations.

This Springer imprint is published by the registered company Springer Nature Singapore Pte Ltd.
The registered company address is: 152 Beach Road, #21-01/04 Gateway East, Singapore 189721, Singapore

Preface

Graphene is a single molecule crystal arrangement of carbon atoms which form a hexagonal honeycomb structure containing sp^2 hybridization bonds and the extended honeycomb network of graphene is the fundamental building block of other essential carbon allotropes. In the 21st century, graphene being the 'wonder' and a 'magical' two-dimensional material has stimulated a lot of research interest and have dramatically shocked the science and technology world due to its fascinating features and unique structure. Moreover, graphene has been discovered to be the strongest and hardest possible material today, hence it can be an efficient alternative for other materials, namely silicon. In the rapid developing arena, graphene has made a remarkable progress in a wide a range of applications particularly in the sectors of biomedicine, composites, energy storage, solar energy, batteries, supercapacitors, chemistry, electronics, physics and many more. Furthermore, due to the evolution of graphene since its discovery in year 2004, therefore, currently it is appropriate to compile all the important fundamental properties, applications and future prospects of this material and this is important. This book consists of 9 chapters that comprehensively covers the recent advances of graphene science, specially focusing on its fabrication, properties and future applications. Starting from the structure of graphene, the basic arrangements are explained in detail. Different forms of carbon are discussed and the allotropy of carbon materials and the performance related to the atomic arrangements are explained. The fabrication of the graphene and graphene oxide materials are illustrated and their specific applications in different fields are depicted. The relationship between the fundamental properties and related applications are clearly explained. Various preparation routes of the graphene and graphene oxide materials are illustrated. Chapter 4 has been prepared exclusively to discuss the different preparation techniques such as chemical vapor deposition, chemical/ mechanical/ electrochemical exfoliation, pyrolysis and arch discharge methods. The resultant materials have been explained in terms of change in structure, shape and other properties. The application of the materials expands to various fields such as batteries, supercapacitors, sensors, LEDs, and water purification, etc. This book will be a good

Introduction

Sachin Sharma Ashok Kumar, Shahid Bashir, Ramesh Kasi, and Ramesh T. Subramaniam

Abstract Since 2004, graphene proved itself a wonder material. Graphene is a two-dimensional layered materials in which layers are single sheets of sp^2 hybrid carbon atoms. Graphene attracted tremendous attention from the researchers around the globe owing to its easy synthesis steps, exceptional properties, and diverse applications. Graphene can be synthesized through many methods. Graphene witnessed its diverse applications because of the unique physico-chemical properties such as large surface area, promising electronic conductivity, strong mechanical strength, extra ordinary electrical and thermal conductivity. due to its numerous outstanding properties, graphene has remarkably created tremendous interest to both physicists and chemist globally with possible potential applications in batteries, sensors, nanoelectronics, flexible displays, supercapacitors, solar cells, organic photovoltaics, hydrogen storage, drug delivery etc.

Keywords Graphene · Two-dimensional · Facile synthesis · Properties · Applications

1 Introduction

Many novel carbon nanomaterials were isolated until Kroto and his coworkers had discovered fullerene in 1985 [1]. After several years, Iijima had discovered carbon nanotubes in 1991 and graphene, a two-dimensional form of graphite, was isolated in 2004 by Andre Geim and Konstantin Novoselov at the University of Manchester, United Kingdom [2, 3]. Graphene is the first ever two-dimensional allotropic form of carbon among other advanced carbon materials which resulted to many new

S. S. Ashok Kumar · S. Bashir · R. Kasi (✉) · R. T. Subramaniam
Department of Physics, Faculty of Science, Centre for Ionics Universiti Malaya, 50603 Kuala Lumpur, Malaysia
e-mail: rameshkasi@um.edu.my

S. Bashir
Higher Institution Centre of Excellence (HICoE), UM Power Energy Dedicated Advanced Centre (UMPEDAC), Level 4, Wisma R&D, Universiti Malaya, Jalan Pantai Baharu, 59990 Kuala Lumpur, Malaysia

© The Author(s), under exclusive license to Springer Nature Singapore Pte Ltd. 2023
R. T. Subramaniam et al. (eds.), *Graphene*, Engineering Materials,
https://doi.org/10.1007/978-981-99-1206-3_1

advancements in research, technology and innovation [4]. In general, the graphene structure is layered and the spacing between each graphene sheet is approximately 0.35 nm [4–6]. The outstanding structural, physicochemical, electronic, and electrochemical features of graphene are remarkably the biggest positive for material scientists [7–9]. Interestingly, graphene is known to have a significantly high electrical conductivity due to the intersection among conduction and valence band at six positions in momentum at the Dirac points. In other words, graphene has been proven to be a near-perfect electronic conductor [4, 10]. In addition, at room temperature, graphene platelets have exhibited a resistivity as low as 10^{-6} Ω cm, thus revealing its high electrical and conductive attributes [11]. Also, superconductivity has been exhibited by the graphene in twisted bilayer form. In addition, having an adsorption power of approximately 2.3% red light and 2.6% green light respectively, the mono-atomic dense bilayer surface of the graphene can be seen by a naked eye [12]. Graphene has also an exceptional clarity for monolayer atomic structure in a vacuum [13]. Previously, it was reported that the high charge room temperature durability of graphene was approximately 15,000 cm^2V^{-1} s^{-1} [14]. To support this statement, some recent researches also reported the identical charge flexibility for hole and electron respectively [9, 15, 16]. Moreover, compared to copper, at room temperature, the splitting stability of charge carriers of audible graphene photons was revealed to be 4.5×10^3 times higher [17]. Graphene has exceptional stiffness and breaking strength having intrinsic strength and modulus of elasticity of 130 GPa and 1 TPa respectively, thus, resulting it to be one of the strongest materials ever [5]. Furthermore, the thermal conductivity of graphene plays a significant role in an efficient and desirable thermal application due to its high ability. For instance, at room temperature, a comparative study was conducted between graphite and graphene, whereby the thermal conductivity exhibited was approximately 2000 W/mK and 5300 W/mK respectively [5, 18]. Illustrated in Table 1 are some of the fascinating physical properties of graphene [4]. In short, due to its numerous outstanding properties, graphene has remarkably created tremendous interest to both physicists and chemist globally with possible potential applications in batteries, sensors, nanoelectronics, flexible displays, supercapacitors, solar cells, organic photovoltaics, hydrogen storage, drug delivery etc.

Table 1 The fascinating properties of graphene [4]

Properties	Graphene
Specific surface area	2630 m^2/g
Young's modulus	–1100 Gpa
Fracture toughness	125 GPa
Mobility charge carrier	2×10^5 cm^2V^{-1} s^{-1}
Thermal conductivity	–5000 W/mK

1.1 Brief Narration of the Book

Since the discovery of graphene, graphene has become a new super-versatile material due to their fascinating properties, thus attracting significant attention from both fundamental and applied research point of view in physics, chemistry, and materials science [19–21]. Today, being known as the thinnest material in the world, graphene is significantly the strongest and most attractive nanomaterial due to their fascinating attributes such as high optical transmittance, high specific surface area, high porosity, high chemical stability, high elasticity, tunable band gap, biocompatibility and ease of chemical functionalization which helps in tuning its properties respectively [4, 5]. In relation to graphene and graphene-based materials, this book demonstrates the impact of the prestigiousness of graphene and graphene-based materials research and gathers all major areas of research and development that has been documented up to date. There are total of nine chapters in this book. Chapter 1 gives a brief overview on graphene, the merits of the Nobel along with a brief narration of the book. Chapter 2 outlines the morphological structure of graphene and other carbon allotropes. In addition, the significance of graphene derivatives is discussed followed by the global statistical analysis of publications and patents related to graphene research, illustrating the advance development of graphene technology. Concisely discussed in Chap. 3 are the 10 different properties of the multifunctional materials. Moreover, the fabrication routes for graphene, starting from the bottom-up synthesis which includes the well-known chemical vapour deposition to top-down synthesis such as mechanical and chemical exfoliation etc. are discussed in Chap. 4. Additionally, Chap. 5 focuses on graphene oxide in terms of their structure, properties, fabrication techniques, characterization, and its future perspectives. Chapter 6 presents the characterization analysis of graphene namely Raman spectroscopy, scanning electron microscopy (SEM), optical microscopy, scanning probe and tunnelling microscopy and transmission electron microscopy and high-resolution TEM respectively. In Chap. 7, the potential applications of graphene are discussed. This is followed by concise sections that highlights 7 major applications of graphene, thus, revealing the versatility of this material in various fields. Chapter 8 discusses about the graphene nanotechnology for renewable energy systems that includes solar and fuel-cell technology, energy storage devices, battery-powered vehicles, and other applications. Finally, Chap. 9 presents the opinions on graphene as a super-versatile material for the near future.

This book is suitable for graphene researchers at all levels and for those who are interested to keep up with the status of graphene when investigating future possibilities. This book covers all aspects and the latest developments of graphene and graphene-based materials along with the latest references ranging from a past decade up to today, thus making this book to be adequate, interesting, and easy to read by all readers.

References

1. Kroto, H.W., Heath, J.R., O'Brien, S.C., Curl, R.F., Smalley, R.E.: Astrophys. J. **314**, 352 (1987)
2. Iijima, S.: Helical microtubules of graphitic carbon. Nature **354**(6348), 56–58 (1991)
3. Novoselov, K.S., et al.: Electric field effect in atomically thin carbon films. Science **306**(5696), 666–669 (2004)
4. Sur, U.K.: Graphene: a rising star on the horizon of materials science. Int. J. Electrochem. **2012**, (2012)
5. Kumar, S.S.A., et al.: New perspectives on Graphene/Graphene oxide based polymer nanocomposites for corrosion applications: the relevance of the Graphene/Polymer barrier coatings. Prog. Org. Coat. **154**, 106215 (2021)
6. Kumar, S.S.A., et al.: Why is graphene an extraordinary material? A review based on a decade of research. Front. Mater. Sci. **16**(2), 1–39 (2022)
7. Simpson, C.D., et al.: Synthesis of a giant 222 carbon graphite sheet. Chem.—Eur. J. **8**(6), 1424–1429 (2002)
8. Geim, A.K.: Graphene: status and prospects. Science **324**(5934), 1530–1534 (2009)
9. Sharif, S., et al.: Introductory Chapter: Introduction to Advanced Carbon Materials and Innovative Engineering Applications. 21st Century Advanced Carbon Materials for Engineering Applications: A Comprehensive Handbook, p. 3 (2021)
10. Jussila, H., et al.: Surface plasmon resonance for characterization of large-area atomic-layer graphene film. Optica **3**(2), 151–158 (2016)
11. Zhu, S.-E., Yuan, S., Janssen, G.: Optical transmittance of multilayer graphene. EPL (Eur. Lett.) **108**(1), 17007 (2014)
12. Nair, R.R., et al.: Fine structure constant defines visual transparency of graphene. Science **320**(5881), 1308–1308 (2008)
13. Kuzmenko, A.B., et al.: Universal optical conductance of graphite. Phys. Rev. Lett. **100**(11), 117401 (2008)
14. Carlsson, J.M.: Buckle or break. Nat. Mater. **6**(11), 801–802 (2007)
15. Novoselov, K.S., et al.: Two-dimensional gas of massless Dirac fermions in graphene. Nature **438**(7065), 197–200 (2005)
16. Morozov, S., et al.: Giant intrinsic carrier mobilities in graphene and its bilayer. Phys. Rev. Lett. **100**(1), 016602 (2008)
17. Neto, A.C., et al.: The electronic properties of graphene. Rev. Mod. Phys. **81**(1), 109 (2009)
18. Balandin, A.A., et al.: Superior thermal conductivity of single-layer graphene. Nano Lett. **8**(3), 902–907 (2008)
19. Kumar, S.S.A., et al.: A comprehensive review: super hydrophobic graphene nanocomposite coatings for underwater and wet applications to enhance corrosion resistance. FlatChem 100326 (2021)
20. Sharma, S.S.A., et al.: The significance of graphene based composite hydrogels as smart materials: a review on the fabrication, properties, and its applications. FlatChem 100352 (2022)
21. Ashok Kumar, S.S., et al.: A review on graphene and its derivatives as the forerunner of the two-dimensional material family for the future. J. Mater. Sci. 1–43 (2022)

Introduction of Graphene: The "Mother" of All Carbon Allotropes

M. Muthuvinayagam, Sachin Sharma Ashok Kumar, K. Ramesh, and S. Ramesh

Abstract Graphene, a two-dimensional (2D), with an electron configuration of $1s^2 2s^2 2p^2$ is a single molecule crystalline structure that forms a hexagonal honeycomb lattice structure. Since its discovery in 2004, graphene has been widely researched for applications in numerous industries namely electronics, optics, automobile etc. The 2D carbon material is known to be the chemical basis for all life on earth, thus, making grapheme potentially an eco-friendly, sustainable solution for various applications. Today, grapheme has been known to be the strongest material due to the presence of the strongest sp^2 C–C bonding. Due to its fascinating properties such as excellent mechanical, thermal, electrical and chemical inertness, graphene is also a suitable candidate for applications associated with batteries and supercapacitors, where by the inclusion of graphene will result in more energy storage in the batteries and supercapacitors.

Keywords Graphene · Graphene oxide · Composites · Properties · Applications

1 Introduction

In the past years, graphene has attracted significant attention globally due to its superior electrical, mechanical, chemical, physical properties and high specific surface area (SSA) respectively [1]. Furthermore, in 2004, graphene was extracted from graphite by Andre Geim and Konstantin Novoselov. In one layer, the carbon atoms were observed to be arranged in a honeycomb crystal structure and the graphene

M. Muthuvinayagam
Department of Physical Sciences, Saveetha School of Engineering, Saveetha University (SIMATS), Chennai, Tamil Nadu 600 077, India
e-mail: muthuvinayaggamm.sse@saveetha.com

S. S. Ashok Kumar · K. Ramesh (✉) · S. Ramesh (✉)
Department of Physics, Faculty of Science, Centre for Ionics University of Malaya, 50603 Kuala Lumpur, Malaysia
e-mail: rameshkasi@um.edu.my

S. Ramesh
e-mail: ramesh@um.edu.my

© The Author(s), under exclusive license to Springer Nature Singapore Pte Ltd. 2023

R. T. Subramaniam et al. (eds.), *Graphene*, Engineering Materials,
https://doi.org/10.1007/978-981-99-1206-3_2

sheets consisted of one atom thick 2D layers held together with strong Van-der Waals forces and with hybridized sp^2carbon atoms. Illustrated in Fig. 1 is the triangular 2D network of a unit cell for a single layer of graphene. For example, consisting of a carbon-carbon bond length of 0.142 nm, the two atoms A & B that are present in the unit cell of single layer of graphene are displaced from one another, thus, resulting each to form a triangular 2D network respectively [2]. Illustrated in Fig. 2 are the other essential types of carbon allotropes [3]. On the other hand, graphite differs from graphene, whereby graphite is formed when many graphene sheets are stacked on top each other and held by Van-der Waals forces. Moreover, when graphene sheets are rolled, this resulted in the formation of another type of carbon allotrope called carbon nanotubes (CNTs) and wrapping the graphene sheets forms the zero-dimensional carbon allotrope, referred to as fullerene.

In a graphene sheet, each atom is connected by a σ-bond and one electron is contributed to a conduction band that extends over the entire sheet. Due to the presence of the conduction band, the graphene gains a semimetal nature with extraordinary electrical properties. Interestingly, heat and electricity very efficiently conducted by graphene and it also strongly absorbs light of all visible wavelengths. However, due to its extreme thinness, a single graphene sheet is nearly transparent. For instance, lately, it was revealed that a single layer of graphene had absorbed 2.3% of incident light in extremely broadband range [4]. In terms of strength, graphene has been reported to be 100 times stronger compared to steel of the same thickness. In short, globally, graphene has been referred to as a 'wonder', thinnest and the strongest material along with its exceptionally high tensile strength, electrical conductivity, thermal conductivity, transparency and many more [1, 2]. Over the years, with regards to the synthesis and applications graphene and its derivatives, the number of patents and publications have increased rapidly and exponentially [5]. As illustrated in Fig. 3, along with its fascinating features, graphene and its derivatives are easily fabricated with various approaches and can be utilized for applications in various fields [6]. For instance, Mohan et al. reported the recent methods for graphene, properties and

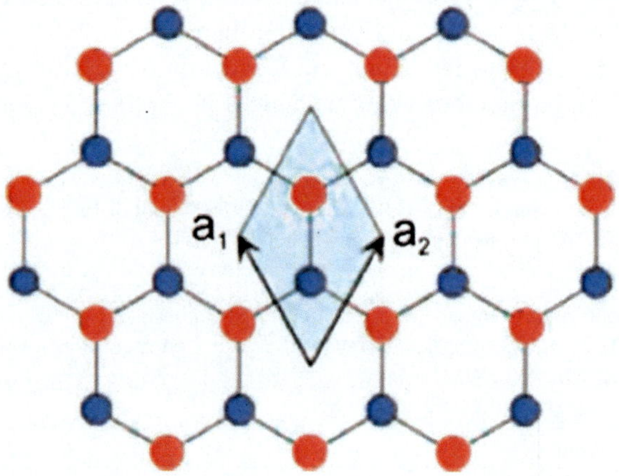

Fig. 1 The honeycomb lattice of graphene. The unit cell defined by vectors a1 and a2 containing the two atoms belonging to sublattices A (blue) and B (red) is highlighted in light blue. Reproduced with permission from Biró et al. [3]

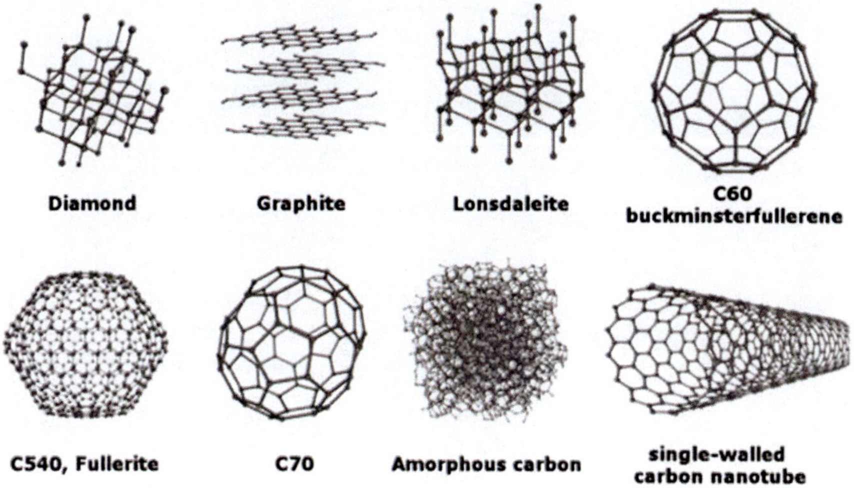

Fig. 2 An overview of different forms of carbon allotropes [https://www.cd-bioparticles.com/support/synthesis-of-carbon-allotropes.html]

their environmental applications, toxicity and safe handling protocols [7]. Furthermore, Wang et al. revealed the latest development in the fabrication and drug delivery application of graphene and graphene-based materials [8]. The bottom-up and top-down synthesis methods to fabricate graphene was also described [9]. Lately, many researchers have devoted significant efforts to fabricate graphene-based electrochemical sensors in order to determine hazardous ions [10]. For instance, from both fundamental and practical perspective point of view, Wu et al. observed the development of graphene-based materials via Raman spectroscopy [11]. Also, Roy and Jaiswal et al. further summarized the applications of graphene-based materials in a wider range of fields that includes drug delivery, nucleic acid delivery, phototherapy, bioimaging and the agnostic respectively [12]. In addition, the fabrication and applications of nanoporous graphene-based materials in various fields were also taken into account [13]. For instance, Phiri et al. reported the graphene production approaches from graphite and some properties and applications in polymer composites [14, 15]. The preparation methods of the graphene and its derivatives were described in detail and the biological and physicochemical characteristics of biomedical relevance were also discussed [16–33].

2 Properties

Graphene has many peculiar properties like high thermal conductivity, electrical conductivity, elasticity, flexibility, hardness and resistance respectively. Graphene has been revealed to not be affected by ionizing radiation. When graphene is exposed

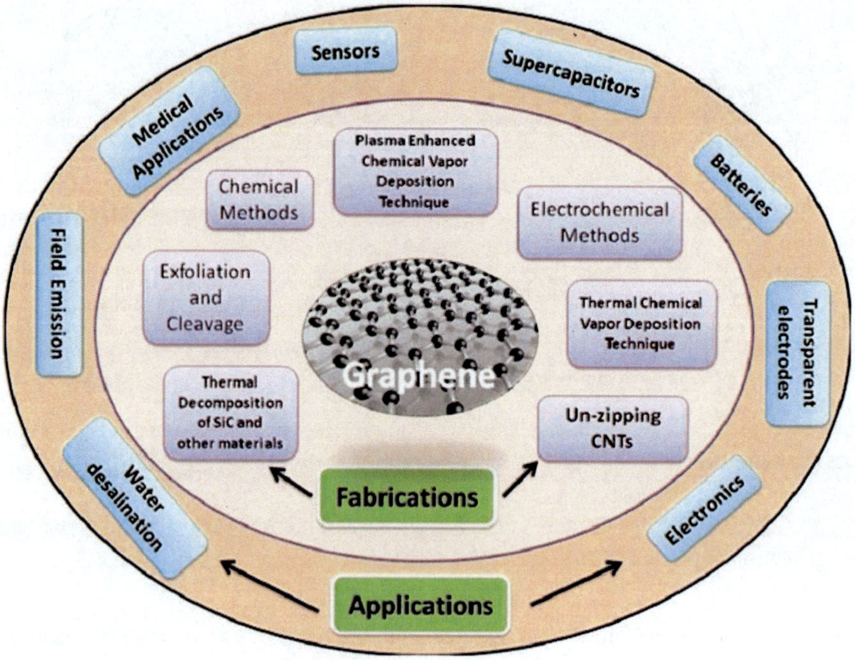

Fig. 3 Schematic model illustrates the versatility of graphene in for numerous applications. Reproduced with permission from Khan et al. [6]

to sunlight, it has the capability to produce electricity. Furthermore, graphene has exhibited a tensile strength and intrinsic strength of approximately 1 TPa and 130 GPa. Furthermore, due to its 2D structure, the SSA of graphene was about 2620 $\frac{m^2}{g}$ [3]. Also, the thermal conductivity that was exhibited by the graphene was relatively high, measured between 4800 and 4300 W/mK [3]. Graphene is also known to be the most stretchable crystal, whereby it can be stretched up to 20% of its initial size without breaking [5–7]. In both experimentally and theoretically, graphene was found to be a near-perfect electronic conductor. For instance, in terms of carrier mobility and electron density properties, graphene has been reported to exhibit high values which were measured at approximately 2×10^5 cm^2/Vs and 2×10^{11} cm^2/Vs respectively [3]. Moreover, in ambient conditions, the mobility and the charge carrier concentration can be tuned and increase up to 1.5×10^4 cm^2/Vs and 10^{13} cm^{-2} [3]. In other words, an ambi-polar electric field effect was displayed by the graphene and between the electron and the holes, and due to having a zero-energy gap, the charge carriers could be continuously tuned. In overall, due to these outstanding properties and its versatility, graphene is a potential candidate for many applications.

3 Graphene Oxide (GO)

GO is made up of different ratios of carbon, oxygen, and hydrogen that are achieved by treating graphite with strong oxidizers and acids. The maximally oxidized bulk product is a yellow solid with carbon and oxygen ratio between 2.1 and 2.9 that retains the structure layer of graphite but with a much larger and irregular spacing. The bulk material spontaneously disperses in basic solutions or can be dispersed by sonication in polar solvents to yield single molecular sheets of GO. In the past years, the GO sheets have been utilized to fabricate membranes, thin films, and composite materials. However, many chemical and structural defects are still associated with the graphene obtained by rGO, thus limiting its uses for some applications but an advantage for a bunch of applications [34–36].

GO, a semiconductor, has exhibited an electrical conductivity between 1- 5 $\times 10^{-3}$ S/cm at a bias voltage of 10 V. Also, the GO disperses readily in water as it has a hydrophilic nature. However, when the suspended GO is treated with hydrazine for one day at 100 °C, a partial reduction of the GO can be achieved. Furthermore, due to the oxidation protocol, the presence of the defects in GO will eventually hinder the efficacy of the reduction. In other words, the quality of grapheme obtained is limited by the precursor quality of the GO and the efficiency of the reducing agent respectively. In general, it was reported that the graphene that has been fabricated from this approach has exhibited an electrical conductivity of below 10 S/cm and a charge mobility in the range of 0.1–10 cm^2/Vs. Moreover, the reported values were found to be much greater than the GO, but lower when compared to pristine graphene, by a few orders of magnitude. Recently, the synthetic protocol for GO with a preserved carbon framework was investigated. Here, it was seen that the overall performance of the rGO were better, exhibiting mobility values of charge carriers of more than 1000 cm^2/Vs for the best quality of flakes.

4 Applications of Graphene and its Derivatives

Graphene is versatile material and therefore it can be incorporated with other elements such as gases and metals to produce various materials and composites with excellent properties. Globally, researchers have widely explored graphene for various applications that include batteries, sensors, LED lights and displays (computer and touch screens), energy generation, supercapacitors, medical and biomedical, water purification, fuel and solar cells respectively. High quality graphene sheets are often utilized in sensors, however, graphene flakes that are produced at low cost and in large scale are frequently adopted in applications that include sports equipment, consumer electronics and automotive etc. [10–13]. In the upcoming sub-sections, the applications of graphene and its derivatives in various sectors will be briefly discussed.

4.1 Batteries

Due to its high electrical conductance, stability, light weight and mechanical flexibility, graphene has shown its versatility in the field of lithium-ion batteries (LIB) [1, 37–40]. However, in an anode material, these properties become more valuable when graphene is employed because graphene provides higher surface to volume ratio, electrical conductivity, surface area and structure adaptability for flexible electrodes which then ensures its strength under extreme conditions. Previously, tin, transition metal and silicon-based anode materials were utilized for LIBs [1, 41]. However, these materials tend to disintegrate after constant charging and discharging, resulting in the overall reduction of the cycling execution. Therefore, to address these challenges, as illustrated in Table 1, graphene is utilized to enhance the performance of the LIBs.

On the other hand, for LIBs and sodium-ion batteries (SIBs) at room temperature, it has been reported that GO is a promising candidate to be utilized as a flexible free-standing battery anode material. Moreover, due to its high surface area, the GO has

Table 1 Summary of graphene-based LIB anode materials

Anode materials	Composition	Discharge capacity for 1st cycle (mAh/g)	Capacity retention after several cycles (mAh/g)	Fabrication techniques	References
Graphene/SnS_2	Hexagonal crystal	1664	After 500 cycles: 600	Solution phase	[42]
Graphene/TiO_2		499	After 10 cycles: 150	Gas/liquid interface reaction	[43]
Graphene/SnO_2		1588	After 40 cycles: 730	Hydrothermal	[44]
Graphene/CuO		640	After 50 cycles: 583.5	N-methyl-2-p yrrolidone Solvent	[45]
Graphene/Co_3O_4		1826	After 40 cycles: 1310	Solvothermal	[46]
Graphene/Fe_3O_4		1426	After 100 cycles: 580	Reduction	[47]
Graphene/Mn_3O_4		900	After 100 cycles: 390	Hydrothermal	[48]

been further investigated as a conducting agent in Li-sulphur battery cathodes. As mentioned previously, the presence of the functional groups in GO can serve as sites for chemical modification and immobilization of active species, whereby, these processes will enable the formation of hybrid structures particularly for electrode materials. Furthermore, it has been revealed that the performance of the batteries was significantly enhanced when the GO-based composites were functionalized with metal oxides and sulfides. Similarly, the functionalized GO-based composites have been widely utilized in supercapacitors, due to its fascinating electronic properties which enables it to escape some of the more frequent restrictions of typical transition metal oxide electrodes. Moreover, researchers are devoting a lot more efforts in this area of research, for instance, by using rGO, as it can remarkably increase the conductivity, and at the same time exploring other approaches that involves nitrogen doping and pH adjustment to enhance the capacitance of the material.

4.2 *Transistors*

In the past years, silicon transistors have been downscaled due to its technological limitations. Since the electronics technologies are evolving drastically, there is a need to enhance the current devices, thus, resulting in more research and development to explore new materials to replace silicon. Recently, remarkable efforts have been devoted by researchers into graphene research due to its fascinating properties particularly the superior electrical properties that include high carrier mobility and high saturation velocity, which are essential for electronic devices. Although graphene was known as a kind of a 3D material, it was not assumed that graphene existed in the free state and was accepted to be flimsy regarding the arrangement of bended structures (e.g., CNTs, fullerene etc.).The graphene channel is synthesized between two gate oxide layers, whereby one is the top gate and second is back gate oxide (substrate) as a layer for the implementation of the dual gate G-FET structure, as illustrated in Fig. 4 [22]. The silicon dioxide acted as the dielectric for the back gate [22]. In addition, Si wafer which acted as the back gate made up the bottom layer. The bilayer graphene channel was synthesized by depositing it onto a thick silicon dioxide layer and was left to grow on a doped Si wafer. When a proper back gate bias voltage was applied, the inversion of the channel was carried out to operate the G-FET as a switch, which can be either turned on or off respectively. The function of back gate was to control the source and drain resistance of the G-FET.

4.3 *Computer Chips*

A new type of computing chip could be a game-changer mainly due its transistors not being made out of silicon. However, it is not as fast or as small as the silicon devices found in current computers or other chip-based devices. However, sooner or later, the

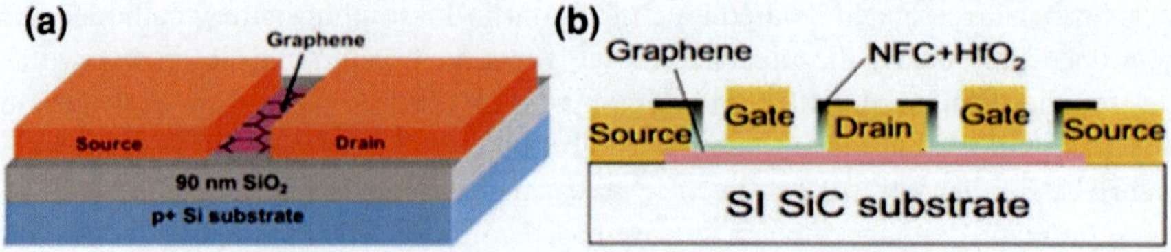

Fig. 4 Schematic model illustrates the **a** back-gated and **b** top-gated GFET. Reproduced with permission from Krsihna et al. [22]

emergence of new computer chips may give rise to electronics that are faster and use less energy. Graphene and its derivatives are being evolved as "miracle materials" and can replace semiconductor in computer chips. However, ultrafast processing computer chips or advanced batteries are not going to become a reality anytime soon [50]. Currently, these strategies are prevented due to some challenges [1]. Firstly, in order to make these systems possible, the graphene has to be completely free of defects and pure to have extremely high conductivity [1]. Currently, it was very unfortunate that no process has been developed to allow the production of large scale of graphene at a low cost. Secondly, a band gap is required by the computer chips in order to function. For instance, the band gap enables the chips to either 1, on or off. Therefore, the conductivity of graphene will be reduced if the band gap is introduced. Hence, enormous research is being conducted to develop a technique that will retain the electrical properties of graphene while enabling it to function as a semiconductor [1].

4.4 Energy Generation and Energy Conversion

Carbon-based polymer nanocomposites are advantageous to be applied in numerous sectors namely the energy storage, aerospace and automotive etc. respectively. In addition, easy processing, configuration adaptability, light-weight and flexibility are some of the essential attributes of these nanostructures. Furthermore, energy storage, fuel cell and supercapacitors have been revealed to be primary components for the future prospective of renewable energy schemes. Moreover, at a low cost, the demand for high energy and power density devices have led to the discovery of nanocomposite materials for various applications that include automotive and electric energy storage devices. For instance, the inclusions of conductive nanofillers (e.g., graphene nanoplatelets, CNTs etc.) into the polymers conductive have proven to enhance the interfacial polarization density of the nanocomposite. Scattered nanocomposites within an insulating pattern performed as nanocapacitor electrodes have resulted in the enhancement of the space charge polarization. Furthermore, it was revealed that the polarization in nanocomposites was possible, specifically at larger frequencies, though the polarization of the conductive nanofiller and the interfacial region may occur at frequencies lower than 1 MHz. On the other hand, the energy storage may need distinctive techniques to store electric power by alternate renewable origins to assure that the devices are suitable enoughto store sufficient energy. In general, the electrochemical storage technique contains the probes, electrolyte media, and power receiver. These components are manufactured from carbon-based materials, metal oxides and materials that are conductive. However, these materials are yet to be explored further and there are great challenges that need to addressed before the devices are launched commercially in the market [25].

Known as an artificial photosynthesis process, by the use of artificial or natural light, the photocatalytic water splitting causes the water to breakdown into hydrogen and oxygen respectively. At present, this method is being significantly explored to produce hydrogen as a clean source of energy. In addition, due to its high surface area and high mobility carrier of electrons, the GO sheets can be incorporated into this process and utilized as a catalyst [35–43]. For instance, the presence of the epoxide and hydroxide functional groups in the GO will allow for more flexible control in the water splitting process. In other words, the band gap and the band positions that are targeted during this process can be tailored by this flexibility. Recently, it has been reported that effective splitting results were exhibited by the photocatalytic activity of GO that contained a band gap within the required limits. Here, effective splitting results were exhibited specifically when a 2:1 ratio (hydroxide: epoxide) was utilized, covering 40–50% of the GO coverage. Interestingly, it has been revealed that GO-based CdS nanocomposites have enhanced the production of hydrogen and the quantum efficiency respectively.

4.5 Supercapacitors

Due to its high-power density, fast charging/discharging rate, long cycle life, a wide operating temperature range and environmentally non-threatening, the supercapacitors have gained tremendous interest around the world. Currently, more work has been focused on the development of different electrode materials such as carbon, conducting polymers and metal oxides for energy storage devices, whereby researchers have devoted more efforts particularly towards the carbon-based materials (activated carbon, CNTs, carbon aerogels). Over a decade, graphene-based composites have been investigated for supercapacitor applications. In general, the specific capacitance of graphene was found to be lesser than the expected value due to restacking of the graphene sheets which can be possibly enhanced by producing a nanocomposite, combined with other materials. For instance, there are many studies that were previously reported related to the synthesis of the transition metal oxides for supercapacitor applications [3, 6]. In short, the nanostructured hybrids constituted by transition metal oxide and graphene are promising for next-generation energy storage devices.

4.6 DNA Sequencing

Due to graphene being extremely thin and having high stability and electrical conductivity characteristics, this have resulted researchers globally to explore the versatility of graphene in medical and biotechnology industries to detect and analyse the DNA sequence, drug delivery, tumor detection etc. In DNA sequencing, the main concept was to create a graphene membrane, which was immersed in a conductive fluid and a voltage was applied to one end so that DNA can be drawn through the graphene's miniscule pores. This method was called nanopore sequencing whereby the DNA can be analysed one nucleotide at a time. This was due to the fact that each nucleotide had different effects on the membrane. O the other hand, various theoretical studies have been conducted in the previous years to improve the DNA sequencing resolution with graphene nanopores [51]. For instance, with the use of Na^+ instead of K^+ ions, the molecular dynamics simulations have revealed that the DNA translocation speed was significantly reduced due to Na^+ ions having stronger interactions [51]. In other words, the interaction strength between the nanopores and the DNA resulted the translocation time to increase exponentially. Also, significant efforts have been devoted to develop nucleotide detection resolution. Here, the molecular dynamics simulations showed that by using graphene nanopores, the base pairs of A-T and G-C could be easily detected [51]. In short, the nanopore sequencing methods are evolving rapidly compared with the other techniques. Due to this rapid advancement, the main goal of sequencing for clinical applications is expected to be achieved in the future. By using nanopores, the single molecule detection has been discovered to be one of the main robust tools for DNA sequencing since it is label free and amplification-free, low material requirement, long read length and high throughput.

4.7 Membrane Separation

In the future, as the population around the world is expected to drastically increase, the high demands for food and water, extended droughts and scarcity of freshwater sources will eventually result in a severe lack of clean water [52]. Therefore, for future energy and water related issues, the cheaper cost and highly-efficient technologies for filtration of alternative water supplies such as wastewater purification etc. and advanced energy storage devices might provide potential solutions [5]. Furthermore, there are other alternative approaches such as distillation and reverse osmosis that could yield potential for high-performance filtering. Previously, CNTs membranes were investigated for water permeation via reverse osmosis towards wastewater treatment and sea water desalination. Interestingly, due to its extraordinary antibacterial properties, the granular activated carbon that contained silicone nanoparticles was revealed to be an effective adsorbent material in commercial water filters [53]. In contrast to the standard routes of storing electrical energy in batteries, through physical ion adsorption/desorption process, the charge is stored by the supercapacitors at the electrode/electrolyte interface, whereas, electric double layer capacitors (EDLCs) store the charge via fast and non-diffusion limited faradaic reactions for pseudocapacitive materials. In other words, due to these features of having fast and highly reversible storage mechanisms, the supercapacitors is a potential candidate for energy storage devices [2].

Alternatively, via the usage of reverse osmosis, the GO was investigated for desalination of water, whereby the GO could be engineered to enable the passage of water and simultaneously retaining some larger ions. For instance, by monolayer or bilayer water, the capillaries with a smaller width generally allows rapid permeation of water. However, in water free conditions, a reasonable mechanical strength is provided by the multilayer laminates. In addition, in humidity free conditions, it is impossible for the Helium to pass through the membranes, but it can easily penetrate when exposed to humidity, however, water vapor passes through without any resistance. Furthermore, when immersed in water, the dry laminates that are vacuum-tight, will act as molecular sieves, thus, blocking some solutes. Additionally, when polycarbonate structure was attached to graphene, it was reported that at initial stage, it was effective to remove salt. As mentioned previously, defects are present in graphene, therefore, using nylon and hafnium metal, the larger and smaller defects can be respectively filled, followed by the addition of an oxide layer, hence restoring the filtration effect. For instance, to filter salty or dirty water, researchers globally successfully developed graphene-based films powered by the sun. Here, in order to fabricate a material that consisted of two layers of nanocellulose, a bacterium was incorporated into the synthesis process. The pristine cellulose was present in the lower layer, whereas, the top layer contained cellulose and GO, which absorbed the energy from the sunlight in order to generate heat. Furthermore, from below, the water is drawn from the system and flows directly into the material. In addition, the water that diffuses into the higher layer, will eventually evaporate, thus, resulting all the contaminations to be left behind. By continuous addition of a fluid coating that hardens, this will produce a membrane film. In short, Bacteria produce nanocellulose fibres along with the inclusion of GO flakes can be produced by the bacteria and this process can be easily manufactured at small or larger scale.

4.8 Coatings

Under dry conditions, the multi-layered films with high optical transmittance are usually produced from GO, due to its permeability attributes. Hence, below a specific size, such films enable the flow of molecules particularly when exposed to water. In addition, these films are made up by millions of randomly stacked flakes, thus, leaving nano-sized capillaries between them. In addition, by chemically reducing it with hydroiodic acid, this will result these nanocapillaries to close and subsequently resulting in the creation of rGO films that are completely impermeable to gases, liquids or strong chemicals (thickness >100). Moreover, for various applications, glass, steel, copper plates etc. coated with graphene can be utilized as containers and the shell life of the medical packaging can be improved by using plastic films that are coated with graphene.

4.9 Lenses

Since its invention, optical lenses have played an essential role in science and technology industry. In addition, with the advancement in micro and nanofabrication methods, continuous compactness of the conventional optical lenses has always played a vital role for a wide range of applications namely data storage, communications, sensors and consumer-driven industries. Moreover, for nano-optics with extremely small structures, there is a great need for lower thickness and smaller sizes of micro lenses, mainly for the visible and near-IR applications.

In the past years, the discovery of the GO along with its superior properties provided novel solutions to overcome the issues associated with the current planar focusing devices. For instance, by dynamically controlling the oxygen content via the direct laser writing (DLW) technique, the giant refractive index modification between the GO and rGO was investigated. Here, it was revealed that the overall lens thickness was significantly reduced (>10 times). Additionally, as the GO was reduced, this resulted in the increment of the linear optical absorption of GO, hence, providing an amplitude modulation mechanism due to the transmission differences between the GO and rGO respectively. Furthermore, over a broad wavelength range, the refractive index and the optical absorption were observed to disperse lesser. In other words, via the DLW technique, a flexible patterning capability was demonstrated by the GO film, which reduced the manufacturing requirements and complexity respectively.

Alternatively, ultrathin planar lens on a GO film was fabricated using the DLW approach. One of the main advantages of the GO flat lenses is that the phase and amplitude modulation can be accomplished at the same time. This was due to the huge refractive index modulation and the variable linear optical absorption of GO that occurred during its reduction process. Furthermore, by varying the lenses size and the laser powers, the focusing intensities and the focal length can be effectively controlled. Moreover, by utilizing an oil immersion high numerical aperture (NA) during the DLW, a 300 nm fabrication feature size on GO film was observed, in other words, the minimum size of the lens was significantly reduced to 4.6 μm in diameter. In other words, by reducing the focal length to as small as 0.8 μm, the NA and the focusing resolution would be potentially increased. In short, the fabrication of GO films with high quality can be easily incorporated on numerous substrates and it can be easily manufactured via the one-step DLW method over a large area at a cheaper cost, thus, resulting the GO-based flat lenses to be a potential candidate for a wide range of practical applications.

4.10 Touch Screens

In the modern electronic technology, touch screen sensors play an important role to grant users to interact with devices. Today, numerous touch screens are widely utilized in consumer electronics such as phones, tablets and other touch-sensitive user interfaces respectively [45]. In general, the sensors are placed in front or embedded in the device's display, thus, enabling direct interaction with the information that is being displayed on the screen. Interestingly, due to their high performance, durability, high sensitivity and multi-touch capabilities, the capacitive sensors are one of the most commonly used touch sensor types. Furthermore, the capacitance variation across the sensor determines the touch location in capacitive sensing. In two layers that are isolated by a dielectric, the electrode arrays are typically arranged, whereby one layer represents the rows and the other columns represents the receiver electrodes respectively. Alternatively, a highly effective LPE method was employed to fabricate graphene using high-speed shear mixing (SM) and tip sonication (TS). For instance, the performance of a combined SM + TS processes for exfoliating graphite in NMP was investigated and at the same time, the results obtained by individual TS and SM processes were respectively compared [26]. Here, the graphene-based touch sensor exhibited a multi-touch functionality of up to four simultaneous touches and a high signal-to-noise ratio (SNR) of approximately 14 dB, while maintaining a high optical transmittance of approximately 78% [27]. In short, it can be concluded that a potential solution can be provided by this approach, concerning issues associated with the standard LPE-graphene production, which demonstrated an effective way to fabricate graphene-based sensors for industrial applications.

5 Conclusions

Graphene is good example of the2D atomic crystals, whereby it is considered to be the mother of all types of carbon allotropes. Due to their exceptional properties, this makes graphene different compared to the 3D crystals. In addition, due to their physical, chemical and optical properties, graphene has attracted significant attention around the world. Overall, as described above, the graphene technology is the fundamental basis for electronics, optoelectronics, biosensing, energy storage, coatings, water purification, hydrogen storage and optical lenses.

References

1. Singh, V., Joung, D., Zhai, L., Das, S., Khondaker, S.I., Seal, S.: Graphene based materials: past, present and future. Prog. Mater. Sci. **56**, 1178–1271 (2011). https://doi.org/10.1016/j.pmatsci.2011.03.003

2. Geim, A.K., Novoselov, K.S.: The rise of graphene. Nat. Mater. **6**, 183–191 (2007). https://doi.org/10.1038/nmat1849
3. Biró, L.P., Nemes-Incze, P., Lambin, P.: Graphene: nanoscale processing and recent applications. Nanoscale **4**(6), 1824–1839 (2012)
4. Bonaccorso, F., et al.: Graphene photonics and optoelectronics. Nat. Photonics **4**(9), 611–622 2010. https://www.nature.com/articles/nphoton.2010.186
5. Zou, L., Wang, L., Wu, Y., Ma, C., Yu, S., Liu, X.: Trends analysis of graphene research and development. J. Data Inf. Sci. **3**, 82–100 (2020). https://doi.org/10.2478/jdis-2018-0005
6. Khan, K., Tareen, A.K., Aslam, M., Zhang, Y., Wang, R., Ouyang, Z., Gou, Z., Zhang, H.: Recent advances in two-dimensional materials and their nanocomposites in sustainable energy conversion applications. R. Soc. Chem. (2019). https://doi.org/10.1039/c9nr05919a
7. Mohan, V.B., Lau, K.T., Hui, D., Bhattacharyya, D.: Graphene-based materials and their composites: a review on production, applications and product limitations. Compos. Part B Eng. **142**, 200–220 (2018). https://doi.org/10.1016/j.compositesb.2018.01.013
8. Wang, Z., Ciacchi, L.C., Wei, G.: Recent advances in the synthesis of graphene-based nanomaterials for controlled drug delivery. Appl. Sci. **7**, (2017). https://doi.org/10.3390/app7111175
9. Bhuyan, M.S.A., Uddin, M.N., Islam, M.M., Bipasha, F.A., Hossain, S.S.: Synthesis of graphene. Int. Nano Lett. **6**, 65–83 (2016). https://doi.org/10.1007/s40089-015-0176-1
10. Molina, J., Cases, F., Moretto, L.M.: Graphene-based materials for the electrochemical determination of hazardous ions. Anal. Chim. Acta. **946**, 9–39 (2016). https://doi.org/10.1016/j.aca.2016.10.019
11. Wu, J.B., Lin, M.L., Cong, X., Liu, H.N., Tan, P.H.: Raman spectroscopy of graphene-based materials and its applications in related devices. Chem. Soc. Rev. **47**, 1822–1873 (2018). https://doi.org/10.1039/c6cs00915h
12. Roy, S., Jaiswal, A.: Graphene-based nanomaterials for theranostic applications (2017). https://doi.org/10.1142/s2424942417500116
13. Wu, X., Mu, F., Zhao, H.: Synthesis and potential applications of nanoporous graphene: a review. Proc. Nat. Res. Soc. **2**, (2018). https://doi.org/10.11605/j.pnrs.201802003
14. Phiri, J., Gane, P., Maloney, T.C.: General overview of graphene: production, properties and application in polymer composites. Mater. Sci. Eng. B Solid-State Mater. Adv. Technol. **215**, 9–28 (2017). https://doi.org/10.1016/j.mseb.2016.10.004
15. Habib, M.R., Liang, T., Yu, X., Pi, X., Liu, Y., Xu, M.: A review of theoretical study of graphene chemical vapor deposition synthesis on metals: nucleation, growth, and the role of hydrogen and oxygen. Rep. Prog. Phys. **81**, (2018). https://doi.org/10.1088/1361-6633/aa9bbf
16. Madni, A., Noreen, S., Maqbool, I., Rehman, F., Batool, A., Kashif, P.M., Rehman, M., Tahir, N., Khan, M.I.: Graphene-based nanocomposites: synthesis and their theranostic applications. J. Drug Target. **26**, 858–883 (2018). https://doi.org/10.1080/1061186X.2018.1437920
17. Martínez, J.I., Moncada, J.L., Larenas, J.M.: The dual descriptor to measure local reactivity on Buckminster fullerenes: an analysis within the framework of conceptual DFT. J. Mol. Model. **16**, 1825–1832 (2010). https://doi.org/10.1007/s00894-009-0638-3
18. Wang, J., Ma, F., Sun, M.: Graphene, hexagonal boron nitride, and their heterostructures: properties and applications. RSC Adv. **7**, 16801–16822 (2017). https://doi.org/10.1039/c7ra00260b
19. Weiss, N.O., Zhou, H., Liao, L., Liu, Y., Jiang, S., Huang, Y., Duan, X.: Graphene: an emerging electronic material. Adv. Mater. **24**, 5782–5825 (2012). https://doi.org/10.1002/adma.201201482
20. Zhang, L., Wang, Y., Niu, Z., Chen, J.: Advanced nanostructured carbon-based materials for rechargeable lithium-sulfur batteries. Carbon N. Y. **141**, 400–416 (2019). https://doi.org/10.1016/j.carbon.2018.09.067
21. Roselin, L.S., Juang, R.S., Hsieh, C.T., Sagadevan, S., Umar, A., Selvin, R. and Hegazy, H.H.: Recent advances and perspectives of carbon-based nanostructures as anode materials for Li-ion batteries. Mater. (Basel) **12**, (2019). https://doi.org/10.3390/ma12081229

22. Krsihna, B.V., Ravi, S., Prakash, M.D.: Recent developments in graphene based field effect transistors. Mater. Today Proc. **45**, 1524–1528 (2021). https://doi.org/10.1016/j.matpr.2020.07.678
23. Siwal, S.S., Zhang, Q., Devi, N., Thakur, V.K.: Carbon-based polymer nanocomposite for high-performance energy storage applications. Polym. (Basel) **12**, 1–31 (2020). https://doi.org/10.3390/polym12030505
24. Natter, N., Kostoglou, N., Koczwara, C., Tampaxis, C., Steriotis, T., Gupta, R., Paris, O., Rebholz, C., Mitterer, C.: Plasma-derived graphene-based materials for water purification and energy storage. C **5**, 16 (2019). https://doi.org/10.3390/c5020016
25. Shin, K.Y., Hong, J.Y., Jang, J.: Micropatterning of graphene sheets by inkjet printing and its wideband dipole-antenna application. Adv. Mater. **23**, 2113–2118 (2011). https://doi.org/10.1002/adma.201100345
26. Tkachev, S., Monteiro, M., Santos, J., Placidi, E., Hassine, M.B., Marques, P., Ferreira, P., Alpuim, P., Capasso, A.: Environmentally friendly graphene inks for touch screen sensors. Adv. Funct. Mater. **31**, 1–15 (2021). https://doi.org/10.1002/adfm.202103287
27. Wang, J., Liang, M., Fang, Y., Qiu, T., Zhang, J., Zhi, L.: Rod-coating: Towards large-area fabrication of uniform reduced graphene oxide films for flexible touch screens. Adv. Mater. **24**, 2874–2878 (2012). https://doi.org/10.1002/adma.201200055
28. Yin, Z., Zhu, J., He, Q., Cao, X., Tan, C., Chen, H., Yan, Q., Zhang, H.: Graphene-based materials for solar cell applications. Adv. Energy Mater. **4**, 1–19 (2014). https://doi.org/10.1002/aenm.201300574
29. Singh, E., Nalwa, H.S.: Stability of graphene-based heterojunction solar cells. RSC Adv. **5**, 73575–73600 (2015). https://doi.org/10.1039/c5ra11771b
30. Choudhuri, P., Bhauriyal, B.: Pathak, recent advances in graphene-like 2D materials for spintronics applications. Chem. Mater. **31**, 8260–8285 (2019). https://doi.org/10.1021/acs.chemmater.9b02243
31. Dragoman, M., Dragoman, D.: The carbon nanotube radio. Proc. Int. Semicond. Conf. CAS **1**, 77–80 (2008). https://doi.org/10.1109/SMICND.2008.4703331
32. Ansari, R., Mahmoudinezhad, E., Motevalli, B.: Different motion patterns of triple-walled carbon nanotube oscillators. J. Vib. Control (JVC) **20**, 773–785 (2014). https://doi.org/10.1177/1077546312455679
33. Strativnov, E.V.: Design of modern reactors for synthesis of thermally expanded graphite. Nanoscale Res. Lett. **10**, (2015). https://doi.org/10.1186/s11671-015-0919-y
34. Karle, N.N.: DFT Study of Adsorption of Trimetallic Endohedral Fullerenes on Graphene (2017)
35. Lu, Z., Ma, L., Tan, J., Wang, H., Ding, X.: Transparent multi-layer Graphene/Polyethylene terephthalate structures with excellent microwave absorption and electromagnetic interference shielding performance. Nanoscale **8**, 16684–16693 (2016). https://doi.org/10.1039/c6nr02619b
36. Akbar, S.: A brief review on graphene applications in rechargeable lithium ion battery electrode materials. Carbon Lett. **28**, 1–8 (2018)
37. Elkholy, A.E., El-Taib Heakal, F., Allam, N.K.: A facile electrosynthesis approach of amorphous Mn-Co-Fe ternary hydroxides as binder-free active electrode materials for high-performance supercapacitors. Electrochim. Acta **296**, 59–68 (2019). https://doi.org/10.1016/j.electacta.2018.11.038
38. Zhao, G., et al.: One-step production of O-N-S co-doped three-dimensional hierarchical porous carbons for high-performance supercapacitors. Nano Energy **47**, 547–555 (2018). https://doi.org/10.1016/j.nanoen.2018.03.016
39. Yu, F., et al.: A zinc bromine "supercapattery" system combining triple functions of capacitive, pseudocapacitive and battery-type charge storage. Mater. Horiz. **7**, 495–503 (2020). https://doi.org/10.1039/C9MH01353A
40. Shao, H., Padmanathan, N., McNulty, D., O'Dwyer, C., Razeeb, K.M.: Cobalt phosphate-based supercapattery as alternative power source for implantable medical devices. ACS Appl. Energy Mater. **2**, 569–578 (2019). https://doi.org/10.1021/acsaem.8b01612

41. Lian, P., Zhu, X., Liang, S., Li, Z., Yang, W., Wang, H.: Large reversible capacity of high quality graphene sheets as an anode material for lithium-ion batteries. Electrochim. Acta **55**, 3909 (2010). https://doi.org/10.1016/j.electacta.2010.02.025
42. Tung, V.C., Allen, M.J., Yang, Y., Kaner, R.B.: High-throughput solution processing of large - scale graphene. Nat. Nanotechnol. **4**, 25 (2009). https://doi.org/10.1038/nnano.2008.329
43. Lu, X., Yu, M., Huang, H., Ruoff, R.S.: Tailoring graphite with the goal of achieving single sheets. Nanotechnology **10**, 269 (1999). https://doi.org/10.1088/0957-4484/10/3/308
44. Wu, Z.S., Ren, W., Wen, L., Gao, L., Zhao, J., Chen, Z., Zhou, G., Li, F., Cheng, H.M.: Graphene anchored with Co_3O_4 nanoparticles as anode of lithium ion batteries with enhanced reversible capacity and cyclic performance. ACS Nano **4**, 3187 (2010). https://doi.org/10.1021/nn100740x
45. Numan, A., et al.: Facile sonochemical synthesis of 2D porous Co_3O_4 nanoflake for supercapattery. J. Alloy. Compd. **819**, 153019 (2020). https://doi.org/10.1016/j.jallcom.2019.153019
46. Kim, S., Zhang, Z., Wang, S., Yang, L., Cairns, E.J., Penner-Hahn, J.E., Deb, A.: Electrochemical and structural investigation of the mechanism of irreversibility in $Li_3V_2(PO_4)_3$ cathodes. J. Phys. Chem. C, **120**, 7005 (2016). https://doi.org/10.1021/acs.jpcc.6b00408
47. Wei, D., Haque, S., Andrew, P., Kivioja, J., Ryhänen, T., Pesquera, A., Centeno, A., Alonso, B., Chuvilin, A., Zurutuza, A.: Ultrathin rechargeable all-solid-state batteries based on monolayer graphene. J. Mater. Chem. A **1**, 3177 (2013). https://doi.org/10.1039/c3ta01183f
48. Colmiais, I., Silva, V., Borme, J., Alpuim, P., Mendes, P.M.: Towards RF graphene devices: a review. FlatChem 100409 (2022)
49. Santra, C.R.: A mini review on graphene-A wonder material for new industrial and biomedical applications. Am. J. Appl. Bio-Technol. Res. **2**(1), 26–29 (2021)
50. Wasfi, A., Awwad, F., Ayesh, A.I.: Graphene-based nanopore approaches for DNA sequencing: a literature review. Biosens. Bioelectron. **119**, 191–203 (2018)
51. Mancosu, N., Snyder, R.L., Kyriakakis, G., Spano, D.: Water scarcity and future challenges for food production. Water **7**(3), 975–992 (2015)
52. Mopoung, S., Moonsri, P., Palas, W., Khumpai, S.: Characterization and properties of activated carbon prepared from tamarind seeds by KOH activation for Fe (III) adsorption from aqueous solution. Sci. World J., (2015)
53. Shao, H., Wu, Y.C., Lin, Z., Taberna, P.L. and Simon, P.: Nanoporous carbon for electrochemical capacitive energy storage. Chem. Soc. Rev. **49**(10), 3005–303 (2020)

Extra Ordinary Properties of Graphene

Maryam Hina, Kashif Kamran, Shahid Bashir, Javed Ahmed, D. Ameer, M. Jahanzaib, and S. Mubarik

Abstract The first two-dimensional atomic crystal is graphene. Before the discovery of graphene and other free-standing two-dimensional atomic crystals, it was widely believed that two-dimensional materials did not exist. Graphene is relatively new but among the most extensively studied materials because of its several promising properties. The distinguishing feature of graphene is the peculiar makeup of its charge carriers. By starting with the Dirac equation rather than the Schrödinger equation, one can more easily and naturally characterise its charge carriers, which resemble relativistic particles. Due to its outstanding qualities, graphene is a strong contender for use in upcoming electrical components such as spin-valve and ultra-sensitive gas sensors, single-electron transistors, and ballistic transistors. This chapter presents the properties of graphene, including mechanical, chemical, thermal, electrical, magnetic, and biological.

Keywords Graphene · Properties · Mechanical · Chemical · Thermal · Electrical · Magnetic · Biological

M. Hina (✉) · J. Ahmed · D. Ameer · M. Jahanzaib · S. Mubarik
Department of Physics, Bahauddin Zakariya University, Multan, Pakistan
e-mail: maryamhina@bzu.edu.pk

K. Kamran
Department of Physics, University of Agriculture, Faisalabad 38040, Pakistan

S. Bashir
Higher Institution Centre of Excellence (HICoE), UM Power Energy Dedicated Advanced Centre (UMPEDAC), Level 4, Wisma R&D, Universiti Malaya, Jalan Pantai Baharu, 59990 Kuala Lumpur, Malaysia

Centre for Ionics Universiti Malaya, Department of Physics, Faculty of Science, Universiti Malaya, 50603 Kuala Lumpur, Malaysia

© The Author(s), under exclusive license to Springer Nature Singapore Pte Ltd. 2023

R. T. Subramaniam et al. (eds.), *Graphene*, Engineering Materials,
https://doi.org/10.1007/978-981-99-1206-3_3

1 Mechanical Properties

Optimal hexagonal crystal lattices are formed by covalently bonding carbon atoms in 2D plane sheets known as pristine graphene structures. Typically, graphene specimens are monolayers that normally adhere to the substrates of SiC or free-standing graphene sheets. A crystalline solid's mechanical properties are often governed by its basic crystal lattice features and structural imperfections such as dislocations and grain boundaries. For instance, the elastic properties of a solid depend on the interactions between the atoms, defect-free crystal lattice and lattice geometry. The presence of crystal defects significantly influences the strength of materials. At the same time, graphene possesses nonlinear elastic behavior and brittle fracture. The graphene's nonlinear elastic response to tensile load is given in Eq. 1:

$$\sigma = \mathrm{E}\varepsilon + \mathrm{D}\varepsilon^2 \tag{1}$$

where σ is the applied stress, ε is the elastic strain, E is Young's modulus, and D is the third-order elastic stiffness.

The graphene's Young's modulus, i.e., 1.0 TPa, which is close to carbon nanotubes, and its third-order elastic stiffness are −2.0 TPa [1]. As a result, the brittle fracture of graphene (σ_{int} = 130 GPa) is the highest ever measured for real materials.

Graphene is highly desirable for structural and other applications due to its exceptionally large values of E and σ_{int}. In addition, graphene can be easily bent. The presence of imperfections critically influences graphene's plastic deformation and fracture characteristics. The experimentally determined imperfections in Graphene include vacancies, Stone-Wales defects, dislocations, and grain boundaries (GBs). Dislocations and GBs cause the most important modifications to graphene's mechanical characteristics. Dislocations, particularly, can function as pathways for plastic flow in graphene, whereas GBs significantly reduce its strength properties.

Due to graphene's extremely high strength, it is used in the synthesis of composite materials to improve their strength and fracture toughness. For instance, adding graphene platelets (GPLs) to ceramic-matrix composites synthesis is more efficient in their toughening than carbon nanotubes and nanofibers [2]. In addition, defect-free graphene possesses the highest Young's modulus (1.0 TPa) and inherent strength (130–130GPa), making it the stiffest substance yet discovered.

There is sp^2 bonding in graphene caused by a honeycomb-like lattice structure. Graphene has the strongest sp^2 orbital hybridisation between P_x and $P_{y,}$ forming a σ-bond. In comparison, the P_z orbital forms a semi filled π-bond, allowing the electrons to move freely. The graphene sheet self-heals by directing the carbon atoms bombarded into vacant sites when exposed to carbon atoms or compounds containing carbon. Faber and Evans demonstrated that graphene has the best ability to deflect the developing cracks, improving its fracture toughness [3]. As indicated in Fig. 1, there are three alternative ways of testing the material's toughness [4].

It was also found that the wrinkling of graphene nanosheets (GNSs) occurs. The internally stored energy within GNSs during wrinkling is insufficient to give the

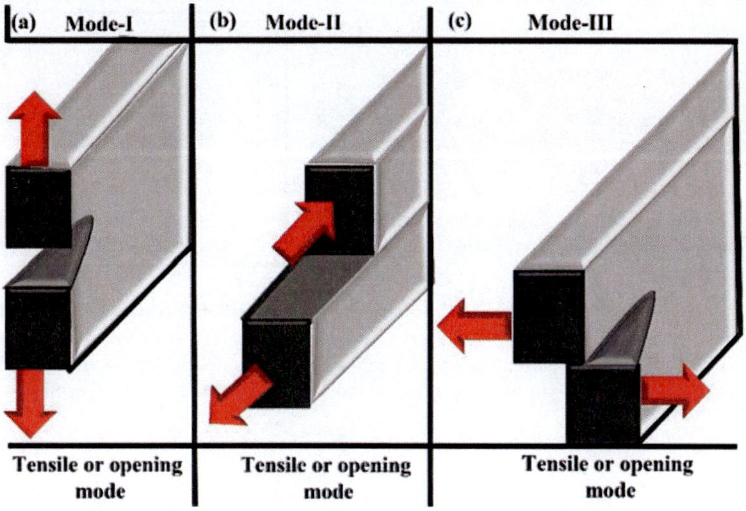

Fig. 1 Modes of material's toughness

GNSs time to regain their shape. On GNSs and exfoliated graphite, wrinkles can be seen. The wrinkling in GNSs is splitting apart at some points while enlarging at others. Wrinkling is irreversible until applying an external force as GNSs cannot store enough elastic strain energy [4].

The research reveals a lack of agreement about graphene's contribution to enhancing other mechanical properties of nanocomposites. For example, some publications claimed that nanocomposites strengthened with GNPs had significantly improved mechanical characteristics [5–9]. But adding GNPs to the epoxy matrix had no appreciable impact on the mechanical properties [10]. Nevertheless, it is demonstrated that GNPs can enhance the mechanical properties of epoxy nanocomposites. Additionally, it was discovered that GNPs enhanced the flexibility of nanocomposites. Covalent functionalised epoxy-graphene nanocomposites were developed by Naebe et al., and an increase of 18% in its elastic strength and 23% in the modulus was reported [10]. An increment of 53% in the elastic modulus of graphene oxide-epoxy nanocomposites was observed [11]. Graphene in epoxy nanocomposites also significantly increases the impact strength and hardness [12]. Qi et al., Lu et al., and Shen et al. synthesised graphene oxide-epoxy nanocomposites with an improved strength of 96%, 100% and 11%, respectively [13–15]. Whereas Bao et al. observed an increase of up to 35% in the hardness of graphene oxide-epoxy nanocomposites [16].

One thing that sets graphene apart from other materials and reinforcing agents in composites are its unique mechanical properties. By utilising AFM to determine the mechanical characteristics of free-standing monolayer graphene (Fig. 2a), graphene was identified "the strongest material ever developed" [1]. The graphene's elastic characteristics and breaking stress reveal that the breaking force depends on the

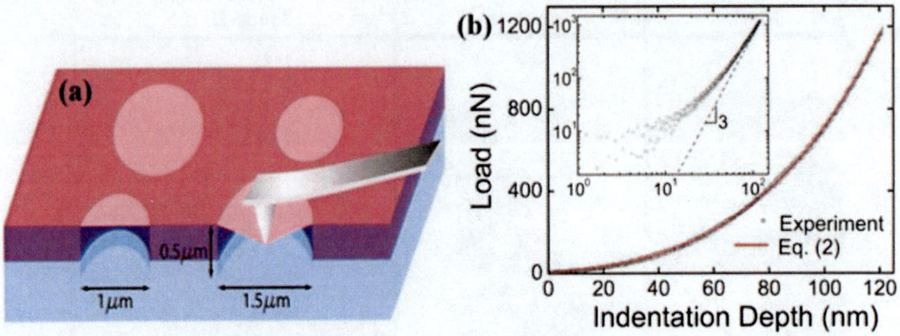

Fig. 2 **a** Illustration of the nanoindentation setup in a suspended monolayer graphene membrane, **b** loading/unloading curve with increasing indentation depth. Reproduced with permission from [17] under CC BY 4.0

radius of tip and is independent of membrane size. Moreover, the force-displacement curves (Fig. 2b) is insensitive to the tip radius.

The intrinsic strength of defect-free monolayer membrane of graphene was calculated to be 42 N m^{-1}, which equals an internal strength of 130 GPa, thus making defect-free monolayer graphene; the strongest material known to man. [1]. Intriguingly, it was shown that graphene's strength and stiffness were preserved even at higher concentrations of sp^3-type defects, with the fracture strength in the sp^3-defect regime just 14% lower than that of pure graphene. However, the strength of the material drastically decreases once it enters the vacancy-defect regime.

The fracture toughness of graphene, a feature highly relevant to engineering applications, is one of the most significant mechanical qualities. To assess the fracture toughness of CVD-produced graphene, Zhang et al. created an in situ micromechanical measurement apparatus and a nanoindenter within a scanning electron microscope [18]. The authors utilised a focused ion beam (FIB) to develop a central fracture in the membrane made of graphene, and when a load was applied, a brittle fracture was seen. The critical strain energy release rate (GC), which was determined to be 15.9 J m^{-2}, and the critical stress intensity factor (KC), which was measured as a critical stress intensity factor (KC) of 4.0 and 0.6 MPa, respectively, were used to determine the fracture toughness of graphene. It is clear from this work and others of a similar nature that the strength of graphene, like that of most membranes, greatly depends on the location of the weakest connection, which is also the point at which failure occurs. Additionally, given the multiple attempts to develop huge graphene sheets at commercial scale, success in this endeavour will presumably produce a material with new flaws and hence less favourable mechanical properties.

One of the more unconventional approaches to exploring and utilising the mechanical properties of graphene is the use of graphene-based shaped layers, which were motivated by Japanese kirigami [19]. Monolayer graphene was worked into springs that could be bent with a magnetic field or stretched up to 240% of their original length without snapping. Future developments could build high-performance mechanical metamaterials based on graphene with microscale dimensions [17].

2 Chemical Properties

A solo sheet of atoms in a 2D honeycomb lattice nanostructure makes up the graphene carbon allotrope. The term originates from "graphite" and the ending "-ene," signifying the abundance of double bonds in the carbon allotrope seen in graphite.

2.1 *Chemical Structure and Bonding*

In a graphene sheet, each atom is coupled to its three closest neighbours via a σ-bond, which increases the size of the conduction band that surrounds the entire sheet by one electron. Graphene is a semimetal with extraordinary electrical capabilities because of these conduction bands, best explained by ideas for massless relativistic particles. Graphene can be used to fabricate field-effect transistors (FET) with bipolar conduction, and its charge carriers have linear, as opposed to quadratic, dependences between energy and momentum. Massive quantum oscillations, enormous and nonlinear diamagnetism, and ballistic charge transfer over very large distances are all visible in the material. As a result, along its plane, graphene exhibits extraordinary electrical and thermal conductivity. Due to its extreme thinness, graphene is almost transparent, yet due to its considerable absorption of light at all visible wavelengths, graphite has a black colour. Furthermore, the material has a nearly 100 times greater strength than the toughest steel of the same thickness [19–21].

Four outer-shell electrons from three atoms engage three sp^2 hybrid orbitals, a grouping of s, p_x, and p_y, orbitals to form σ-bonds. These bonds have an average length of 0.142 nm. The last outer-shell electron fills p_z orbital parallel to the plane. In response to these orbital hybridisations, two partially-filled bands of movable free electrons are generated, capable of tuning most of graphene's remarkable electrical characteristics (Fig. 3).

Graphene sheets stack to form graphite at a distance of 3.35 nm between them. In solid graphene sheets, diffraction patterns often show graphite's (002) stacking. Certain single-walled nanostructures display this characteristic. For example, presolar graphite onions include Graphene that has only (hk0) rings and is not stacked.

2.2 *Geometry and Stability*

Transmission electron microscopy (TEM) of graphene sheets hung between metal bars shows the isolated graphene's hexagonal lattice structure in a single layer. A flat sheet appeared to be "rippling" in several of these pictures, with an amplitude of roughly one nanometer. Due to the instability of two-dimensional crystals, these ripples may be a property of the substance, or they may result from the continual filth

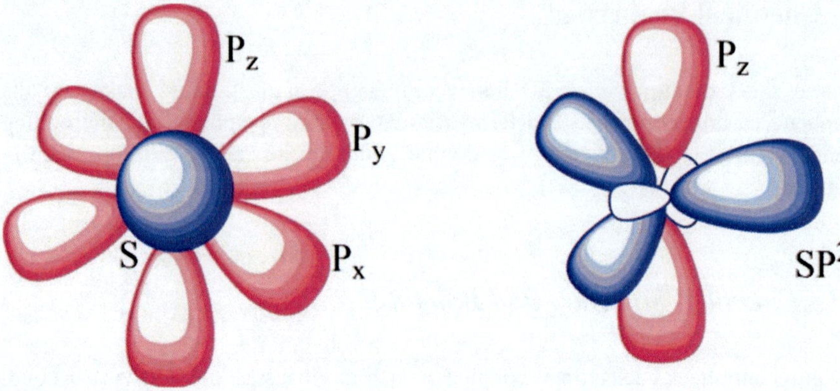

Fig. 3 The hybrid orbital sp^2 with three major lobes at 120° is formed from the carbonorbitals 2 s, $2p_x$, and $2p_y$. The last orbital, pz, is protruding from the graphene plane

seen in all TEM pictures of graphene. To get atomic-resolution images, photoresist remnants that appear as "adsorbates" in TEM images may need to be removed. This would account for the apparent rippling. Figure 4 depicts the π and σ bond in graphene.

Sigma (σ) bonds are created when sp^2 hybrid orbitals overlap, whereas π bonds are created when projecting p_z orbitals tunnel together. (Fig. 5) mounted on silicon dioxide substrates also show the hexagonal configuration Scanning tunnelling microscopy (STM) pictures of Graphene. The ripples visible in these photos are not intrinsic; rather, they are a product of the substrate's lattice's conformity to the graphene, which causes them.

Fig. 4 π and σ bonds in Graphene

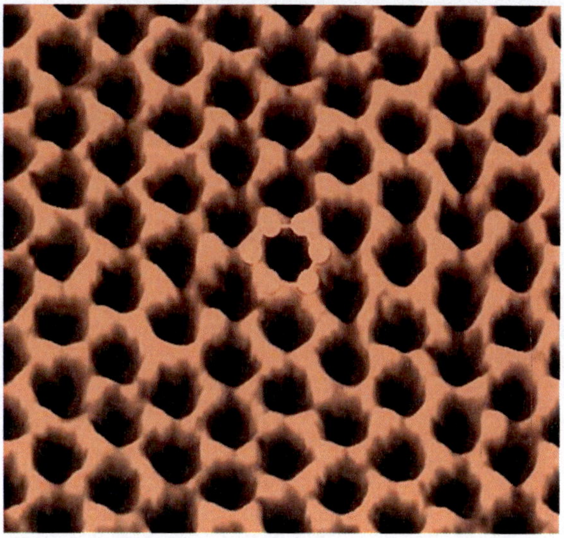

Fig. 5 Scanning probe microscopy image of Graphene

2.3 *Chemistry of Graphene Materials*

Among its neighbours, graphene is the substance that is the thinnest. It has a high irreversible nature of 130 GPa, a high Young's modulus of 1 TPa, and strong heat conductivity as a single sheet. It also possesses significant quantum-war influence characteristics. It illustrates the impact of the ambipolar electric field using section voltage. The chemical makeup and structural characteristics of graphene are discussed. As a standalone substance, graphene has no qualities. Still, once it is broken down and transformed into composites like GO, rGO, fGO, and others, it acquires the features needed in various real-world technological and biomedical applications. To produce graphene, micromechanical fracture, anodic coupling, optical abrasion, liquid phase exfoliation, processing silicon carbide, and many other sequential procedures are required. It shows the extraction procedure for graphene. Hung used gel electrophoresis to test the GO on the outside surfaces to back up composites' mechanical characteristics. The process is much quicker and easier. Using GO in composites, such as fibre-braced polymers, is inconceivable. GO can improve the composites' interfacial holding properties by successfully encasing the polymer's carbon fibre and carbon surfaces. The process of turning graphene into fibre enhanced the material's retaining capacity. However, the structure could be overpriced because every fibre must be treated [19–24]. Since graphene typically does not react with oxygen, various techniques are employed to synthesise graphene oxide. The topping and bottoming cycle of the graphene oxide production process is depicted in Fig. 6.

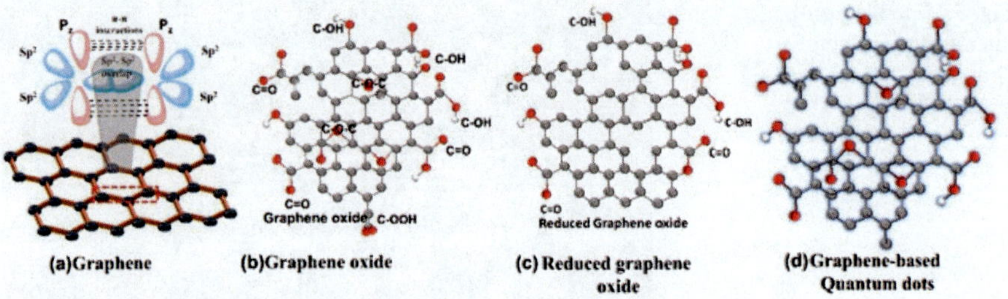

Fig. 6 **a** Pristine graphene and chemically modified graphene, **b** Graphene Oxide with –CO and –OH functional groups, **c** Reduced Graphene Oxide with only –OH functional groups and **d** Graphene Quantum Dot (GQD)

3 Thermal Properties

Because of the unique technique, the carbon atoms in graphene and related compounds are bound together, these materials offer particularly intriguing thermal properties. This section looks at graphene's specific heat and thermal conductivity and similar materials. We also look at the conditions that must be met to create a ballistic heat flow that does not scatter.

3.1 Phonon Dispersion of Graphene

Examining the material's lattice vibrational modes (phonons) is the first step toward comprehending graphene's thermal characteristics. N = 2 carbon atoms make up the graphene unit cell, shown in Fig. 7 by dashed lines.

The relationship between the phonon energy E and the phonon wave vector q is known as dispersion. Transverse (T) modes refer to in-plane displacements perpendicular to the wave propagation direction, whereas longitudinal (L) modes relate to atomic displacements along the wave propagation direction (shear waves). Transverse modes with two equivalent polarisations can exist in normal three-dimensional (3D) materials. However, graphene's special two-dimensional (2D) structure lets flexural (Z) phonons move atoms out of a plane. Examining the material's lattice vibrational modes (phonons) is the first step toward comprehending graphene's thermal characteristics.

The change in energy density U caused by a 1 K change in temperature is represented by a material's specific heat C as given in Eq. 2.

$$C = \frac{dU}{dT} \tag{2}$$

T is the temperature in absolute terms. When referring to heat capacity or specific heat, the units of joules per kelvin per unit mass, per unit volume, or mole are

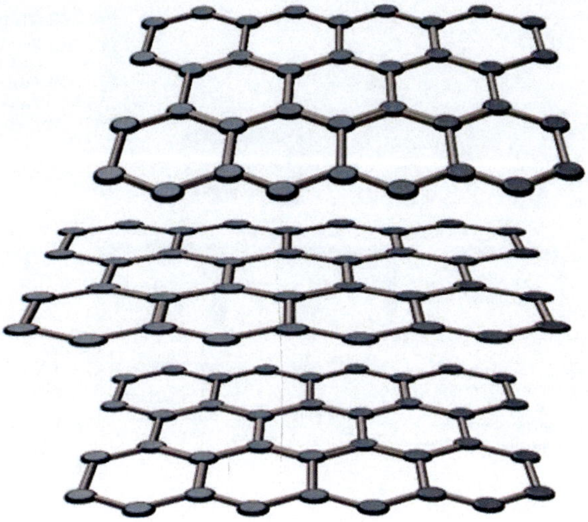

Fig. 7 Atomic arrangement in graphene sheets

sometimes used interchangeably. The thermal time constant (τ = RCV) of a body, where R is the thermal resistance for heat dissipation (the inverse of conductance, R = 1/G), and V is the volume of the body, is determined by the specific heat in addition to the thermal energy that is stored inside it. The lattice vibrations (phonons) and free conduction electrons of a material store specific heat.

$$C = Cp + Ce \quad (3)$$

However, at all significant temperatures (>1 K), phonons predominate the specific heat of graphene, and the phonon-specific heat is directly related to temperature, as depicted in Fig. 8. The specific heat is virtually constant at extremely high temperatures (nearing the in-plane Debye temperature of 2100 K) as given in Eq. 4.

$$Cp = 3NAKB \approx 25Jmol - 1K - 1 \approx 2.1Jg - 1K - 1 \quad (4)$$

The Boltzmann constant, KB, and Avogadro's number, NA, are used in the aforementioned Dulong-Petit limit. Graphite's specific heat at ambient temperature is Cp ≈ 0.7 J/g K, or around one-third of the theoretical top limit [20, 21]. However, due to the poor coupling between graphite layers, this value for graphite at normal temperature is around 30% greater than that for diamond [20]. Similar behaviour is anticipated for thermally excited isolated graphene sheets at ambient temperature [22].

Guo and his co-workers proved that the edge effect causes a decrease in thermal Transport with an increase in thermal conductivity. On the other hand, the energy gap between different phonons also decreases with dimmer lines increment [23]. After that, Pop et al. determined that thermal conductivity is sensitive to the edge shapes, widths, and strains of graphene nanoribbons (GNRs) [22]. Therefore, the probability

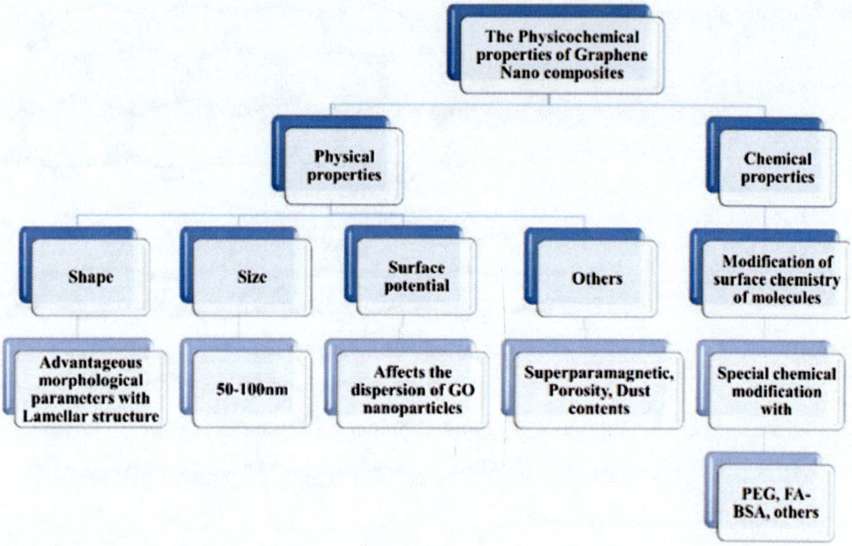

Fig. 8 Physicochemical properties of graphene

of phonons Umklapp process will increase, and thermal conductivity will reduce [24]. Furthermore, when three or more waves interact in a solid, such as electron waves or lattice waves, the umklapp process occurs. As a result, the sum of the wave vectors equals a vector in the reciprocal lattice rather than zero [25]. Combined with their unique electronic properties, graphene may be suitable for future nanodevices [23].

The six Dirac points in momentum space are where the conduction and valence bands of the semimetal graphene meet. Its Brillouin zone's vertices are split into two different groups of three points that are not equal. The labels for the two sets are K and K'. The sets result in a valley degeneracy of $gv = 2$ for graphene. In contrast, conventional semiconductors' main region of interest is typical, where momentum is zero [25]. Its four electrical characteristics set it apart from other condensed matter systems.

I. Electron transport

Graphene exhibits exceptional electron mobility at normal temperature, with measured values of 15000 cm^2V^{-1} s^{-1} [25]. Therefore, it was anticipated that the mobilities of holes and electrons would be similar.

II. Electron energy spectra

Applying the tight-binding approach to the two inequivalent atoms in the graphene unit cell while considering their nearest four neighbours yields the electron dispersion curve for graphene, which is then used to determine the electron-phonon scattering rates. With ($k_x = 0$, $k_y = 0$) being the point or centre of the Brillouin zone, this method gives the following dispersion relation for the electron conduction band in two-dimensional Graphene as a function of momentum along the k_x and k_y axes (Eq. 5) [26].

$$E(k_x, k_y) = \gamma \sqrt{1 + 4\cos\left(\frac{\sqrt{3}k_x a}{2}\right)\cos\left(\frac{k_y a}{2}\right) + 4\cos^2\left(\frac{k_y a}{2}\right)} \quad (5)$$

Between 10 and 100 K, the mobility is almost completely independent of temperature, suggesting that defect scattering is the main scattering mode. Acoustic phonon scattering in graphene restricts its mobility at ambient temperature to 200000 cm^2V^{-1} s^{-1} at a carrier density of 10^{12} cm^{-2}, 10 × 10^6 times larger than copper [26].

III. Electrical resistance and resistivity

Graphene sheets have an equivalent resistivity of about 106 cm, which is lower than the resistivity of silver. On SiO_2 substrates, however, optical phonons have a greater impact on the scattering of electrons than graphene's phonons. This has a 40000 cm^2V^{-1} s^{-1} mobility limit [26]. When electrons in copper come into contact with impurities, resistance goes up in a way that is proportional to length [27]. Two kinds of transportation are most common. One is thermally activated, while the other is ballistic and temperature independent. Ballistic electrons are similar to carbon nanotubes, which are spherical. Resistance goes up sharply at room temperature at a certain length, 16 μm for the ballistic mode and 160 nm for the other mode. Even at room temperature, graphene electrons can travel micrometres without scattering.

IV. Electrical conductivity

Although, there is no carrier density close to the Dirac points, graphene has a minimum conductivity. It is unclear where this minimum conductivity came from but localised puddles of carriers that permit conduction can be produced by wavy regions in the graphene sheet or ionised impurities in the SiO_2 substrate. In addition, the concentration of contaminants affects the conductivity. In conclusion, controlled potassium doping of clean-graphene devices in UHV at low temperatures has shown that the conductivity of graphene depends on the density of charged impurities [26]. Graphene is also utilised as a supplementary conductive component to improve the conductivity of the metal-component structures. Doping, which can raise carrier densities, has been investigated as an alternate method to increase graphene's conductivity. However, it does not directly address the negative effects of line defects on conductivity. Doped graphene films frequently have a low degree of stability, as seen by those whose graphene sheet resistance increases by around 40% in a few days [27].

4 Physico-Chemical Properties

Different graphene-based materials, including graphene (hydrogenated graphene), fluorographene (fluorinated graphene), graphdiyne, porous graphene, graphene nanoribbon (GNR), graphene oxide (GO), and reduced graphene oxide, have been created using synthetic methods (rGO) [27]. Figure 8 depicts a few physicochemical features of Graphene. Figure 9a shows an atomically thin sheet (single layer) of carbon atoms with sp^2 bonds organised in hexagonal patterned graphene. Single-layer graphene (SLG) exhibits exceptional characteristics, such as a zero-band gap because the p and p* bands touch the Dirac point. Furthermore, SLG electrons behave like massless fermions at the Dirac point, as in Fig. 9b [28].

According to Fig. 10a, carbon is the sixth element in the periodic table, and its ground-state electronic configuration is $1s^2 2s^2 2P_x{}^1 2P_y{}^1 2P_z{}^0$. While the energy level $2p_z$ is similar to $2p_x$ and $2p_y$, it is retained without an electron for convenience. As shown in Fig. 10b, the nucleus of a carbon atom is surrounded by six electrons, four of which are valence electrons. Three types of hybridisation—sp, sp^2, and sp^3—can be created by these electrons in the valence shell of a carbon atom. For example, the development of sp^2 hybrids is seen in Fig. 10c. Carbon atoms create a layer of honeycomb network with a planar structure, commonly known as monolayer graphene, when they share sp^2 electrons with their three adjacent carbon atoms. The parallelogram in Fig. 10d represents the unit cell of a graphene crystal. The lattice constant of the unit-cell vectors a1 and a2 is 2.46.Å, and two carbon atoms make up each unit cell. Considering that the electrons can move and resonate makes the planar ring stable [29].

In an ordinary sp^2 hybridisation of adjacent neighbouring carbon atoms on the graphene layer, an out-of-plane π bond is composed of $2p_z$ orbitals perpendicular to the planar structure (Fig. 10d). In contrast, an in-plane σ bond comprises sp^2 (2 s,

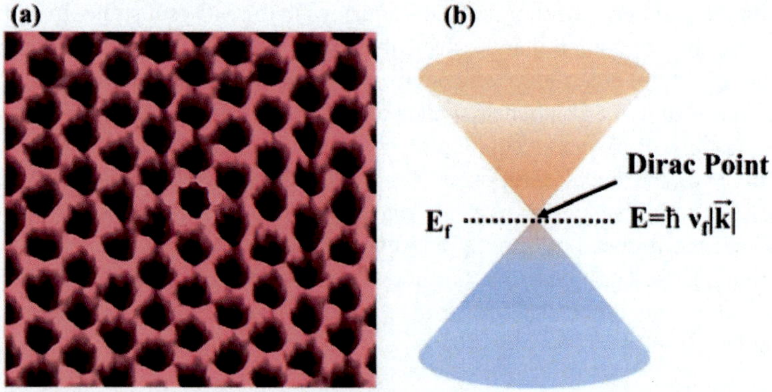

Fig. 9 Schematic illustration of the graphene **a** Single-layer graphene, **b** Conduction and valence bands touching at the Dirac point in the energy–momentum diagram for one of the discrete spots of the graphene's Brillouin zone [28]

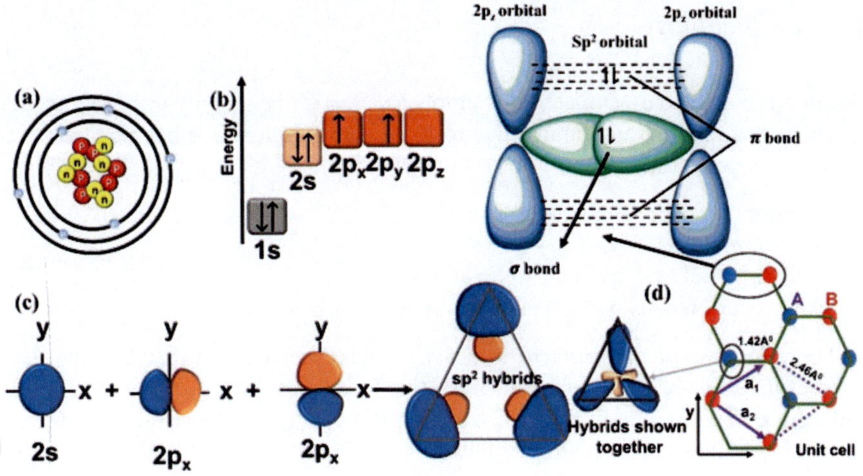

Fig. 10 **a** An atom of carbon's atomic composition **b** Energy levels of carbon atoms' outer electrons **c** sp^2 hybrids formation **d** Graphene crystal lattice; A and B are carbon atoms belonging to different sub-lattices, a1 and a2 are unit-cell vectors. **e** σ bond and π bond are constructed by sp^2 hybridisation

$2p_x$, and $2p_y$) hybridised orbitals. The covalent σ bond resulting from this has a short interatomic length of 1.42 Å and is tougher than the sp^3 hybridised carbon-carbon bonds in diamonds, giving monolayer graphene its exceptional mechanical characteristics. In monolayer graphene, the conduction band (CB) and valence band (VB) with a zero band gap form from the half-filled band that permits free-moving electrons. The -bonds also offer a weak van der Waals link between nearby graphene [30].

The molecular scale of nanomaterials affects a variety of distinctive features. Because a nanomolecule's size or diameter is in the nanometer range, it has a high surface area ratio and the quantum size effect, which causes the nano molecules to have significantly different properties from other common materials [26]. From a toxicological standpoint, particle size is one of the key physicochemical characteristics that affect the toxicity of graphene family nanoparticles (GFNs). Due to an increase in surface area and more opportunities for cellular contact, a reduction in size to the nanoscale promotes greater cellular absorption. According to numerous studies, the size of the nanoparticles impacts the process and effectiveness of cellular absorption, circulation, distribution, clearance, and toxicity of GFNs. Particles with diameters smaller than 40 nm can enter the nucleus, particles smaller than 100 nm can enter the cell, and particles smaller than 35 nm can pass the blood-brain barrier [30].

4.1 Permeability

The thinnest available 2D material is graphene. Despite being only one atom thick, graphene has fascinating features due to its impermeability. Graphene is the most impermeable material due to:

(a) its ultrathin (thinnest) structure
(b) high strength
(c) graphene's closed-spaced carbon atoms, and
(d) high electron density repels all molecules.

The p-orbitals of graphene form a dense, delocalised cloud that fills the gap between its aromatic rings. As a result, repelling field develops and prevents hydrogen and helium-like tiniest molecules from passing through, even when a pressure difference is 1–5 atm across their atomic thickness at ambient temperature. The ability of graphene to maintain its structural integrity at high pressure is due to its high fracture strength of 42 N/m and Young's modulus of 1TPa. In comparison, theoretical studies agreed with experimental findings. According to theoretical calculations, no gap in electron density around aromatic rings would block molecules from passing. The geometric gap calculated from the van der Waals is smaller than the size of He. Because the C–C bond length in graphene is 0.142 nm, the pore size based on the nucleus alone is 0.246 nm. Graphene can be regarded as the most impermeable material because no molecule can pass through it and because the permeance of a membrane rises as the membrane's thickness decreases. Graphene is also the thinnest 2D material, making it the most impermeable substance (Fig. 11).

Impermeable graphene can also be utilised as a barrier membrane for harmful environmental substances and as an anti-corrosion covering. In numerous chemical reactions, graphene is an inert substance that permits other substances to react. For example, on graphene-coated nickel and copper, current densities drastically decrease during anodic and cathodic scanning, reducing reaction rates or ionic transfer. This suggests that the metal layer beneath the graphene serves as a shield. A similar line of thought says that graphene with controlled holes can be used to make the perfect thin membrane for filtering fluids or gases. Theoretically, the artificial pores in graphene can make it more porous and let only certain things through. Since atomically thick graphene is an impermeable membrane, a perfect sheet of graphene between two aqueous reservoirs of ions will not let ions move from one reservoir to the other because it is not permeable.

In addition, graphene properties can be combined with its impermeability to improve its characteristics further. For instance, corrugated surfaces can be coated with a conformal coating to make them impenetrable using graphene's elastic and flexible carbon-carbon bonds. Using the high transparency and conformability of electrons, it is possible to make atomically thin, impermeable liquid cells that can be used to take pictures with a transmission electron microscope (TEM). Owing to graphene's strong electrical conductivity and resistance to high-pressure variations, liquid can be retained, and an electrical charge is minimised during electron

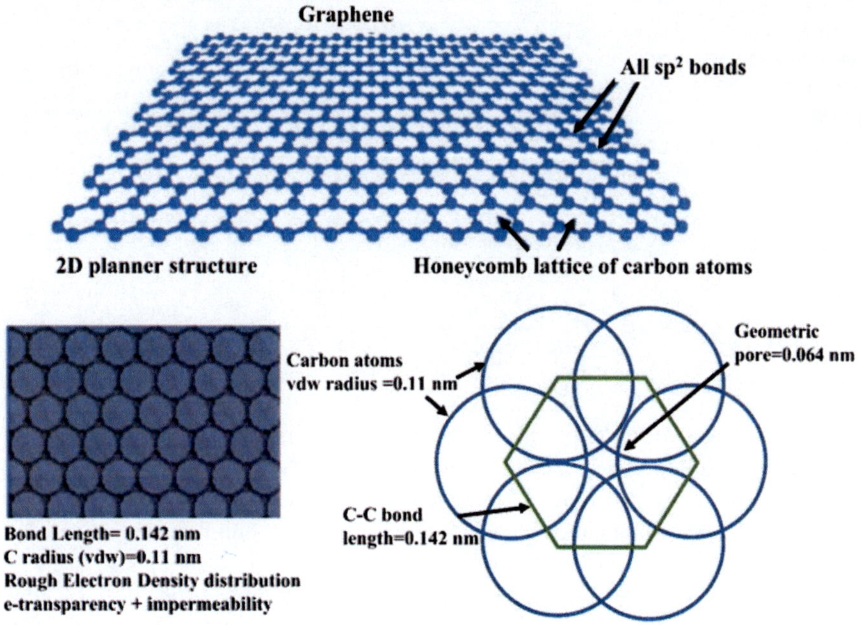

Fig. 11 Sp^2 hybridised carbon atoms are organised in a 2D honeycomb lattice to form the graphene lattice (Bottom). The molecular structure with a rough distribution of the electronic densities: While graphene is essentially impenetrable to all molecules at ambient temperature, it is reasonably transparent to electrons. Additionally, the geometric pore's size (0.064 nm) prevents molecules from passing [31]

microscope imaging. Graphene-membrane chemical interfaces can also be used as a protective layer to stop metal from rusting, control how much water can pass through, and keep dangerous nanomaterials inside [31].

Bunch et al. compared a micro-chamber capped with a graphene sheet and silica to test the permeance of several gases and concluded that the permeation occurs through the microchamber walls rather than the graphene sheet [32]. The pressure change is used to calculate the leak rate values using Eq. 6.

$$\frac{dN}{dt} = \frac{V}{k\beta T}\frac{dP}{dt} \tag{6}$$

Despite being only one atom thick, defect-free graphene is considered completely impermeable to all gases and liquids. This conclusion is supported by experiments that showed that gas permeation through micrometre-sized membranes could not be detected within a detection limit of 10^5–10^6 atoms per second [33]. Furthermore, perfect graphene sheets have recently been shown to be impermeable to standard gases, including helium. Graphene has new uses since it is a very thin but impermeable film. The local density approximation (LDA) and the general gradient approximation

Table 1 The energy barrier height of an atom through perfect and defective graphene [34]

Defect	Local density approximation (LDA)	General gradient approximation (GGA)
No defect	18.77	11.69
Stone Wales defect	9.21	6.12
Di-vacancy	8.77	5.75
Tri-vacancy	4.61	3.35
Tetra-vacancy	1.20	1.04
Hexa-vacancy	0.37	0.44
Deca-vacancy	0.05	0.10

(GGA) were used to figure out the energy barrier height of an atom going through both perfect and imperfect graphene. This is shown in the Table 1.

It can be observed from Fig. 12 that the magnitude of the defects, as determined by the number of carbon atoms involved in their creation, diminishes exponentially with the height of the penetration barrier.

Mallineni et al. looked at how N-dopants affected the impermeability of few-layered graphene (FLG) made by chemical vapour deposition on copper [35]. Because the grain boundaries in FLG do not offer a continuous path for the transportation of gas due to its extreme tortuosity, they have a minor impact on their oxygen permeability. However, they demonstrated that multiple N-dopant combinations in FLG lead to structural flaws that let gas molecules through selectively. They investigated the effects of dopant configuration on graphene impermeability using various techniques. Non-graphitic nitrogen dopants make holes in graphene, letting oxygen through and oxidising the Cu substrate. Graphitic nitrogen dopants, on the other hand, do not change from their natural state. Additionally, they discovered that altering the dopant arrangement effectively allows one to tailor the work function of graphene (Fig. 13) [35].

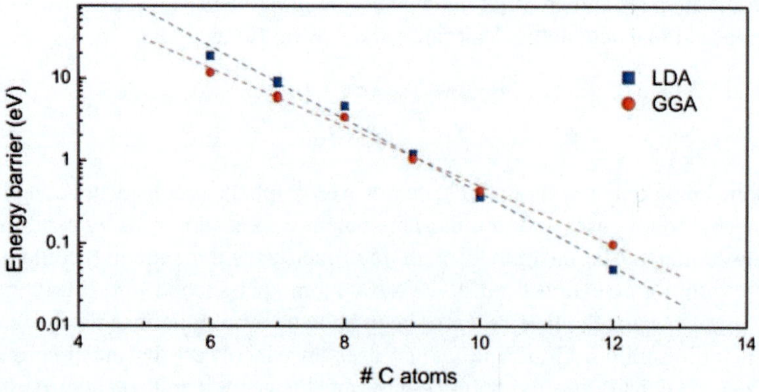

Fig. 12 Energy barrier v/s no. of missing atoms. Reproduced with permission from [34]

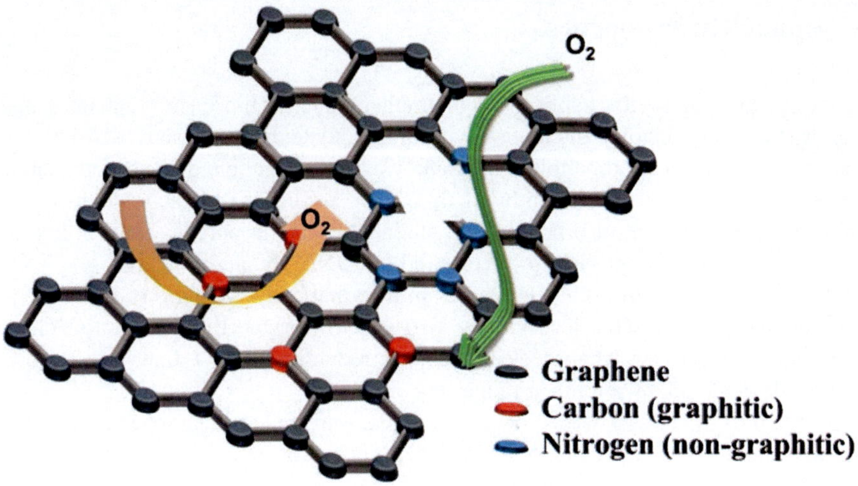

Fig. 13 Influence of dopants on the impermeability of graphene

The strongest chemical bonds in nature make up graphene, which also has certain other unusual features. Graphene has been researched as an excellent membrane that stops organisms from travelling from one side to another since changing these connections require a lot of energy. Even though graphene mono-layers are only one atom thick, tests have shown that they can stop noble gases from getting through, trap bacteria, and stop metals from rusting when used as a protective coating [36].

The barrier qualities of graphene and the chemically derived graphene oxide (GO) membranes are diverse. Without any flaws, monolayer graphene is impenetrable to all gases and liquids. It also possesses excellent physical and chemical stability and low toxicity, just like graphite. These characteristics are expected to set graphene apart from other barrier materials in the market. The graphene films created via chemical vapour deposition (CVD) are widespread with flaws and grain boundaries, and instead of shielding copper from oxidation, they hasten the corrosion of that metal. This issue might be resolved by using multilayers made up of graphene. In this sense, GO is particularly attractive since multilayer films may be produced quickly and affordably by spraying GO solutions onto various substrates, coating objects with them using a dip or rod method, etc. It is demonstrated that the permeation characteristics of the generated GO laminates are quite exceptional. Under the dry state, they are impenetrable to even helium, but in humid conditions, they offer little protection against water vapour. When submerged in water, the laminates act as molecular sieves, permitting tiny ions to pass through while obstructing bigger ones [37].

5 Biological Properties

Given that most biological applications of graphene, such as biosensors, cell imaging, and drug delivery, involve interaction with blood, it stands to reason that this material should be blood compatible. Graphene was initially applied in sensor-related fields without direct contact with living things. According to one theory, the single-stranded DNA quickly adsorbed onto graphene, producing potent chemical bonds that enhanced its receptivity to corresponding DNA. In addition, the graphene surface successfully prevented DNase from enzymatically cleaving ss-DNA [38]. The HRP/ss-DNA/GP/GC electrode was also used to fabricate third-generation electrochemical biosensors. It had an electrocatalytic reduction of H_2O_2 that worked well, was stable, and had a wide linear range.

A recent study showed that a biocompatible scaffold made with graphene improves the differentiation of human mesenchymal stem cells into bone cells while preventing their uncontrolled multiplication. The promise of graphene for stem cell research was demonstrated by the differentiation rate, comparable to the one attained with conventional growth agents. It has been reported that biocompatible nanographene oxides of various physical sizes have been covalently grafted with polyethylene glycol star polymers onto the chemically activated surfaces and edges, giving the NGO aqueous stability in buffer solutions and other biological environments. The visible to near-infrared (NIR) photoluminescence seen with these NGO was employed for cellular imaging. Doxorubicin was loaded onto an NGO with a high capacity and was selectively carried into particular cancer cells by antibody-guided targeting, pointing to the potential uses of graphene materials in biology and medicine. It has been demonstrated that adding graphene considerably enhanced chitosan's modulus, even at extremely low concentrations, and the composite was well-biocompatible with L929 cells. Macroscopic antibacterial graphene-based paper with superior bacterial growth suppression capability is easily fabricable in another paper. This might be helpful in some eco-friendly applications. However, a relatively recent study found that graphene oxides had dose-dependent toxicity to cells and mice, causing lung granuloma development that cannot be removed by the kidney and promoting cell apoptosis [39].

Much progress has been made in studying graphene for medication delivery using in vitro tests. However, the in vivo behaviour of graphene loaded with pharmaceuticals has to be studied for clinical cancer and other disease treatment. Liu and colleagues first investigated PEGylated GO photothermal treatment and in vivo tumour uptake utilising xenograft mice models for the first time [43]. Due to the highly effective tumour passive targeting of GO brought on by the EPR phenomenon, they saw a very high tumour uptake of the PEG-modified GO. Furthermore, a highly effective tumour killing was accomplished using substantial GO absorbance in the near-infrared (NIR) region in conjunction with low-power near-infrared (NIR) laser irradiation of the tumour. In addition, strong antibacterial effects can be seen in GO and rGO papers. This discovery creates new prospects for the use of GO in environmental and medical applications due to the scalability and low cost of the graphene-based antibacterial paper [43].

By limiting the actions of bioactive pharmaceuticals to the desired areas only, drug delivery systems (DDS) have been developed to reduce their adverse effects. These systems have also been employed to extend the duration of medical interventions. Creating a targeted drug delivery system, also known as a smart delivery system for identified targets, and a sustained and responsive release system for the pharmaceuticals are the current difficulties in DDS research. Controlling intricate transport and surface phenomena will be necessary to realise an ideal release system (e.g., diffusion, degradation, swelling, release profiles, and adsorption of DDS elements).

Because of its excellent functionality and versatility, we predict that graphene will have a wide range of prospects as a DDS carrier in this context. For example, one effective tactical approach is the physical bonding of hydrophobic, van der Waals, or stacking interactions to attach water-insoluble hydrophobic bioactive compounds to the surface of graphene. In addition, attaching water-soluble molecules to graphenes containing bioagents can be further altered to become soluble in aqueous solutions. The resultant graphene compounds may aid the medication's general efficacy [44].

Neuronal self-organisation is mediated in a developing nervous system by coordinating several mechanical, chemical, and electrical signals. The lack of non-invasive techniques to measure the process is the reason for understanding the problem of the spatiotemporal regulation of this emerging network, particularly in humans. A novel kind of microelectrode is now being created using graphene to analyse the complex brain circuitry. As cardiac pacemakers or peripheral nervous system stimulators, graphene microelectrodes may find greater use. Due to graphene's non-magnetic and anti-corrosive characteristics, these probes may also lengthen the lifespan of neural implants.

Conversely, unlike metallic implants, the non-magnetic behaviour of graphene enables the secure, artefact-free reading of magnetic resonance images. In addition, graphene's inherent low noise properties are crucial for high signal-to-noise ratio brain circuit recording. The flexibility of graphene is another benefit; as a result, thin, flexible electrodes can be created to surround neuronal tissue. Recently, they also found that the work function of graphene may be efficiently tuned by modifying the dopant arrangement. Transparent graphene-based neural micro-electrodes may enable optogenetic control of the underlying neural circuitry and simultaneous optical imaging and electrophysiological recordings.

Additionally, graphene has high biocompatibility and significantly encourages mouse hippocampus cell neurite sprouting and outgrowth. Finally, the observed prevention of amyloid fibrillation by GO sheets and their protein-coated surfaces raises the possibility that graphene can remove amyloid monomers and prevent neurodegenerative disorders [40].

Graphene has the potential to be used as an antimicrobial coating for surgical equipment or other surfaces since it may kill bacteria, inhibit the growth of corrosive and pathogenic microorganisms, and prevent their occurrence. After being exposed to graphene materials, bacterial viability declines. One of the two potential pathways is the physical harm that frequently results when bacterial membranes come into contact with the ragged graphene sheet edges. Additionally, oxidative stress appears to be at play. Smaller particle sizes and oxidised graphene materials seem to have stronger antibacterial properties [41].

Proteolytic enzymes frequently hamper the process of delivering drugs in the cytoplasm. GO is a carrier for effective gene and medication delivery (Fig. 14). GO may conjugate with various polymers and biomolecules thanks to its functional group (COOH and OH) (ligand, DNA, protein). One method is to functionalise it with a cationic polymer, such PEI. It might have a significant interaction with DNA and RNA's negatively charged phosphate ions, making it useful as a non-viral gene vector. In addition, it increases cell selectivity, decreases cell toxicity, and facilitates effective transfection.

Due to the special characteristics of graphene and its capacity due to the ability to generate conjugates with other functional groups, it is possible to produce biocomposites with altered properties. Materials based on graphene are essential for the engineering of bone and tissue. Reconstructing bone defects depend heavily on the mineralisation and regeneration of bone tissues. Since spontaneous regeneration is not always feasible, one of the most popular techniques is using grafts. Synthetic bone substitutes based on ceramic are employed in clinical practice due to the high cost of grafting. Some drawbacks include the deficiency of sustainable cells, the possibility of immunogenicity, and the transmission of infection. Graphene and its derivatives create materials with improved mechanical and osteogenic qualities when mixed with ceramics. For improved capabilities, graphene has been mixed with HAp, a widely used material [42].

Numerous sensing investigations, including detecting mercury ions, DNA, and certain genes, have already been conducted using graphene-based nanomaterials. Graphene-based nanomaterials can facilitate enzymatic catalysis or act as enzymes

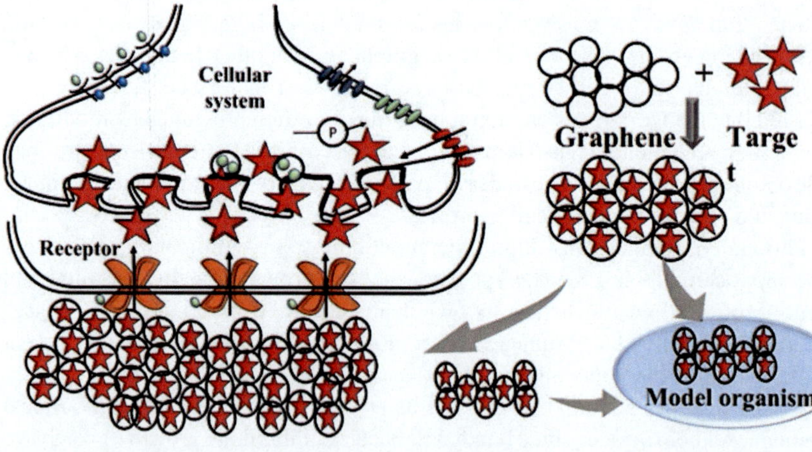

Fig. 14 Graphene as a carrier for targeting gene or small molecular drug delivery

in simulated reactions, particularly useful in applying biosensors featuring enzymatic catalysis. For example, the reactions of glucose oxidase can be enhanced using graphene-based materials (GOx), horseradish peroxidase (HRP), cytochrome, laccase, and bilirubin oxidase [43]. When detecting preclinical disease, graphene-based nanomaterials have been a conventional primary prevention strategy to measure and track metabolic disorders [44].

Physical and chemical action modes are part of graphene's antibacterial mechanism. The most frequent types of damage are physical and are brought on by the sharp edge of graphene coming into close contact with bacterial membranes. Additionally, photothermal ablation and wrapping are part of the physical damage mechanism. Reactive oxygen species and charge transfer-induced oxidative stress are linked to the molecular mechanisms of action (ROS).

Several antibiotics and antibacterial enzymes have attacked different bacterial infections. However, the regulated and localised deployment is essential for delivering a secure and increased effect at therapeutic concentrations, given their obvious toxicities when administered intravenously. Due to its huge specific surface area and electron-rich surface, graphene has been widely explored as an emerging drug delivery technology. Therefore, graphene could be functionalised with antibiotics and antibacterial enzymes for their prolonged delivery and increased antibacterial action by merging the benefits of drug loading and intrinsic antibacterial activity. Vancomycin, levofloxacin, cefalexin, ciprofloxacin, and enzymes like lysozyme have been functionalised using graphene-based materials recently and have been discovered to provide increased antibacterial ability.

Van, a glycol-peptide antibiotic, attaches to the cell wall of Gram-positive bacteria through D-alanyl-D-alanine moieties, which may prevent it from working properly during the transglycosylase step of peptidoglycan production. This causes the cell wall stiffness to decrease, killing the bacterial cells. Additionally, because of the interactions between graphene and Van, which allow for continuous release after initial fast release, the latter could provide long-lasting pathogen suppression in addition to the natural antibacterial activity of graphene [45]. Table 2 shows graphene's pros and cons with relevant materials for biological applications.

As a result, graphene and its connected materials have emerged as key producers of cutting-edge tools for biological applications. The properties of graphene and its associated materials constitute a state-of-the-art tool for creating new, highly compatible biomaterials that may pave the way for a revolutionary drug administration and regenerative medicine method. The core of every decision-making process continues to be weighing production methods against cytotoxicity and performance, and graphene materials are taking up more and more place in the world of cutting-edge medicine.

We are adamant that high controllable and good electrical characteristics are a starting point for graphene's employment in biological applications. These elements would boost its competitiveness and pave the way for a successful future for graphene materials [46].

Table 2 Graphene and similar materials' benefits and drawbacks for use in biological applications

	Benefits	Drawbacks
Graphene	• High conductivities, both electrical and thermal • High functionalisation control	• Hydrophobicity • High cost • Difficult workability • Small production
GO	• Water dispersibility • Polar functionalization • Low cost • Easy workability	• Lower electrical and thermal conductivity • Surface random functionalisation • Poor control on post-preparation functionalization
rGO	• High electrical and thermal conductivities • Good control on functionalisation • Low cost than pristine Graphene	• Hydrophobicity • Difficult workability • Properties related to production methodology used

6 Physical Properties

The Geim group published one of the first papers on graphene in 2004 that proved the viability of an electric field effect in the material, i.e., the capacity to change the carrier density in the graphene sheet by simply applying a gate voltage [47]. For instance, graphite, charcoal, and diamond are some examples of the allotropes of the element carbon. Graphene is an allotrope of carbon. However, despite this, it has unique features and physical traits and frequently breaks records in these areas.

One of graphene's most surprising characteristics is that its charge carriers behave like massless relativistic particles, or Dirac fermions, and can move with minimum dispersion in the surrounding environment. This peculiar behaviour has caused numerous uncommon events in graphene. The valence and conduction bands of graphene barely overlap, making it a zero-bandgap 2D semiconductor. The charge carrier densities of up to 10^{13} cm^{-2} and room temperature mobilities of ~10000 cm^{-2} s^{-1} are measured due to their significant ambipolar electric field influence. By altering the chemical potential using the electric field effect, it has been discovered that graphene exhibits an uncommon half-integer quantum Hall effect (QHE) for both electron and hole carriers.

Additionally, graphene has a high degree of transparency, with visible light absorption of only 2.3%. For a single-layer sheet at room temperature, its thermal conductivity, k, is calculated with a 5000 W mK^{-1}. Moreover, graphene has exceptional mechanical strength [48]. The maximum electron mobility of any material, or the rate at which electrons may move through it, is found in graphene. Since graphene's electron mobility is thought to be 100 times greater than that of silicon, it is a particularly desirable material for application in electronic devices. The incredibly high carrier mobility has drawn the greatest attention. There have been reports of mobilities of more than 100 000 cm^2/V.s and saturation velocities of around 5×107 cm/s.

In the world of technology, graphene is used under specific conditions and as a component of more complicated construction. For instance, different scattering interactions can affect electrical transport. These include scattering caused by close-range interactions with neutral defects or adsorbates, roughness, and phonons, as well as long-range interactions with charged impurities on graphene or, more likely, on the supporting insulator substrate.

The features of the environment in which the graphene occurs and the quality of the graphene itself determine which mechanism dominates the scattering. For instance, when graphene is in contact with polar substrates like SiO_2 or Al_2O_3, Coulomb scattering from charged impurities often predominates at low temperatures. Scattering is also present when the substrate is removed and the graphene is hung because of the absorbates. High mobilities are possible because phonon scattering takes over when this graphene is heated, and the adsorbates are volatilised. When the graphene is flawed, carrier transport will be dominated by scattering from neutral point defects. The magnitude of the carrier mobility and its relationship to temperature and carrier density could be used to determine the sort of scatterer that predominates in a certain graphene sample (n). Therefore, mobilities above 100 000 $cm^2/V.s$ suggest acoustic phonons are the dominant kind of scattering, where AC 1/nT. Long-range Coulomb scattering results in mobilities of 1000–10,000 $cm^2/V.s$, independent of n. Neutral faults become significant in extremely faulty samples or at high carrier densities and SR1/n [49].

The heat conductivity of graphene is ten times greater than that of copper, making it unique material with the highest thermal conductivity. Since it provides an essential tool for the thermal management of electronic components, this trait may also be advantageous for electronic applications.

The material's flexibility, mechanical toughness, and thinness, as well as its high thermal conductivity (up to 5000 W/mK) and extremely high current carrying capacity (up to $10^9 A/cm^2$) [49].

The virtual transparency of graphene is an additional quality that adds to its lengthy list of outstanding qualities and could be useful for electrical applications. When you consider that graphene is only one atomic layer of carbon atoms in its monolayer form, its transparency becomes clear. Approximately 97.7% of visible light can pass through graphene in this form. Graphene is a very intriguing choice for electronic applications in displays or solar cells, among many other things, because of its conductivity and transparency. Engineered roperties of graphene and its applications are given in Table 3.

The virtual transparency of graphene is an additional quality that adds to its lengthy list of outstanding qualities and could be useful for electrical applications. When you consider that graphene is only one atomic layer of carbon atoms in its monolayer form, its transparency becomes clear. Approximately 97.7% of visible light can pass through graphene in this form. Graphene is a very intriguing choice for electronic applications in displays or solar cells, among many other things, because of its conductivity and transparency [50].

Due to its outstanding qualities, including its high surface area, inherent mobility, Young's modulus, electrical conductivity, and thermal conductivity, graphene has

Table 3 Properties of graphene and its applications

Engineered properties	Applications
High room temperature and ballistic electron transfer with mobility up to 200000 cm^2v^{-1} s^{-1}	• High speed transistors • Spin devices • Single electron transistor • Semiconductor memory • Quantum hole resistance standard • Silicon replacement • MEMS
The high strength of 1100 gpa modulus with fracture strength of 30gpa	• Composite materials • Pressure sensors
High curability due to low density ~2 g/cm^3)	• Wiring materials
High surface to weight ratio	• Energy storage such as fuel cells
High light transparency	• Transparent electrodes and laser materials
High sensitivity for chemicals; thus, physical properties can be chemically tuned	• Chemical and biosensors • Hydrogen storage materials
High barrier materials due to high impermeability if defect-free	• Coatings and membranes

received the most attention among carbon-based materials. Additional graphene derivatives that are in high demand for electrodes include graphene oxide (GO), reduced graphene oxide (rGO), graphene hydrogel, graphene aerogel, graphene quantum dots (GQDs), GO quantum dots (GOQDs), and doped graphene. This is particularly true given their advantageous electrochemical properties, high surface area, lightweight, and wide range of electrical conductivity values. Future technologies will need materials based on graphene to function as either primary components or additives because they have shown sufficiently remarkable features at the laboratory scale. With the use of graphene-based materials, various applications, including those in the environmental, structural, electrical, and optoelectronic industries, and energy storage in particular, are showing significant advancements and are frequently approaching production levels. As anticipated given the variety of morphologies and materials [51]. Physical properties and their relevant device applications of GO and RGO materials are given in Fig. 15.

Improved dispersibility in typical organic solvents, a crucial step in creating innovative graphene-based nanocomposites, is one of the primary goals of pure graphene sheet functionalisation. Furthermore, functionalisation is expected to affect intrinsic traits, including electronic properties, to enable the regulation of conductivity and band gap for cutting-edge nanoelectronics devices. According to Georgakilas et al., there are commonly two approaches for organic covalent functionalisation reactions employing graphene and GO [52].

1. Covalent bond formation between dienophiles or free radicals with the pure graphene's C = C bonds.
2. GO's oxygenated aliphatic domains and organic functional groups form covalent bonds.

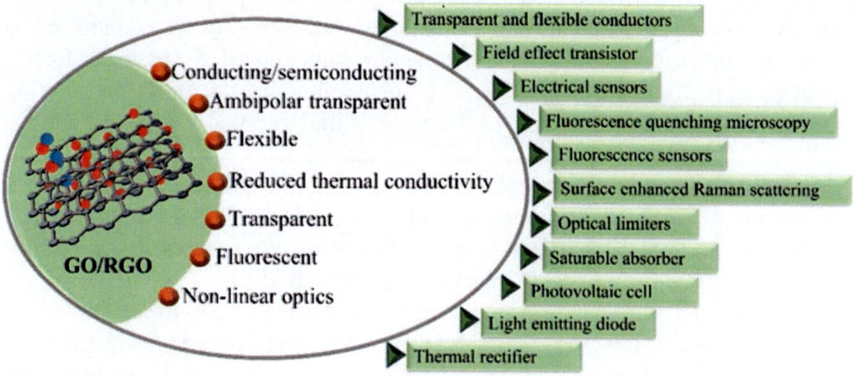

Fig. 15 Physical properties and their relevant device applications of GO and RGO materials

Most graphene nanocomposites are made by making nanoparticles right on top of the graphene scaffold or by putting nanoparticles together on the surface of the graphene through covalent or non-covalent interactions. The nanoparticle-decorated graphene has unique and improved physical properties that canot be reached by either component alone. These properties have a lot of potential for use in optoelectronic materials, biology, and other fields (Fig. 16) [52, 53].

Devices may also be made using graphene's mechanical characteristics. For instance, Bunch et al. claim to have developed the "world's thinnest balloon," which

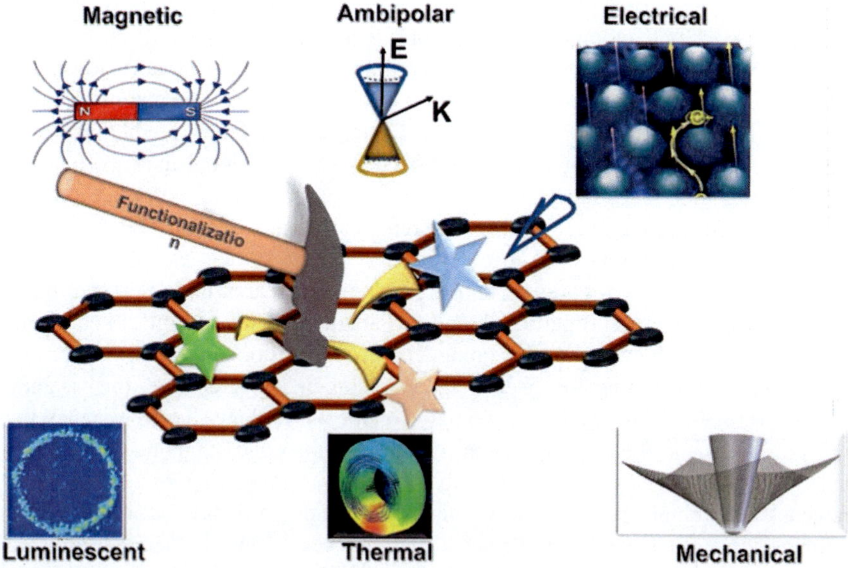

Fig. 16 Exceptional properties of functionalised graphene

is impermeable to gases [32]. They postulate that this property may be utilised as membrane sensors for pressure changes in small volumes, as selective barriers for gas filtration, as a platform for photographing graphene-fluid interfaces, and for creating a physical barrier between two phases of matter. Similar to this, Stolyarova and his group showed that gaseous bubbles could be captured and later controlled by an AFM tip between a graphene monolayer and the surface of a SiO_2 substrate. Devices with "lab-on-a-chip" technology may use this in the future [54]. Some of graphene's strange properties have been tried to be copied in a 2D electron gas with high mobility modulation confined in an AlGaAs/GaAs quantum well.

This hexagonal superlattice is projected to exhibit electronic dispersion with a pseudospin degree of freedom similar to Dirac's. Moreover, because the parameters can be changed, making these systems artificially seems better than making them out of natural graphene. A standard 2D electron gas constrained in semiconductor devices nanopatterned with a honeycomb shape would offer exceptional chances to explore the physics of Dirac-fermions [50].

7 Magnetic Properties

For modern industry, magnetic materials are crucial. These are often ferromagnets at room temperature and make up most of the currently employed magnetic materials. These elements are 3d- or 4f-transition metals like Fe, Co, and Ni. The partially filled d- or f-electron bands are where magnetic ordering originates. Since regulating the spin of s- or p-electrons will significantly expand the study boundaries of magnetism, the development of transition-metal non-involved magnets has long piqued attention. Since their structures are often stable, straightforward, adaptable, and simple to modify, carbon-based materials are particularly interesting because they make spin induction more possible and theoretical magnetism prediction easier.

Furthermore, the bonding type in graphene is so exceptional that it offers the opportunity to simultaneously generate localised spins and couple them when altering these bonds in specific modes, leading to magnetic ordering. In a nutshell, graphene-based materials are thought to be potential for use as s- or p-electron magnets. It has been assumed by theoretical and experimental research that graphene might possess certain magnetic properties, including Para magnetism, spin-glass behaviour, and magnetic switching events (ferromagnetic or antiferromagnetic). There are two techniques for generating localised magnetic moments in graphene: the vacancy approach (creation of the magnetic moments on the basal plane sites by vacancy via ion irradiation) and the edge approach (creation of the edge magnetic moments at the edge sites by edge-type defects). Both the vacancy and edge approaches have the drawback of creating just a small amount of magnetism at the vacancy or edge regions of the graphene sheet. Because they are susceptible to being passivated by their environment, the vacancy and edge magnetic moments are unstable. In contrast to chemical doping, which can maintain graphene's structural stability while introducing additional point defects, high-density vacancies can weaken and cause it to lose its structural stability. N-doping is a reliable way to introduce magnetic moments into graphene, and different types of N atoms contribute to distinct localised magnetic moments, according to theoretical research. Because N adatoms are highly stable, the edge magnetic moments they produce are also stable.

Graphene quantum dots (GQDs) are graphene nanoscale structures with outstanding photoluminescence capabilities due to their strong quantum features and edge effects. Only Curie-like Para magnetism has been observed in studies on graphene oxide quantum dots (GQDs) produced by annealing. The graphene nanoscale structures known as graphene quantum dots (GQDs) have strong quantum properties and edge effects that provide exceptional photoluminescence capabilities. Since the GQDs produced by thermal annealing allow for both edge detection and reconstruction, it is assumed that most spin polarisation at the edge states has been switched off [55].

The ability of covalently functionalised graphene with sp^3-type defects makes it magnetic everywhere. Tuning graphene' electrical or magnetic properties is difficult because of the material's chemical stability, making it difficult to regulate exact sp^3-functionalization. Three following additional strategies can make it better [56]:

1. Hydrogen-doping approach
2. Flourine-doping approach
3. Hydroxyl-doping approach.

Graphene samples with a layer count between 2 and 7, produced using various techniques, also show the presence of strong ferromagnetic properties and antiferromagnetic traits akin to frustrated or phase-separated systems. The preparation of graphene samples included thermal exfoliation of graphitic oxide (EG), conversion of nanodiamond (DG), arc evaporation of graphite in hydrogen (HG), and reduction of single-layer graphene oxide with hydrazine hydrate (EG-H). Magnetic hysteresis at room temperature is seen in all of the graphene samples. The number of layers and sample area impact the magnetic characteristics of graphene samples, with low values for either favouring higher magnetism. The magnetic characteristics of graphene are influenced by molecular charge transfer, with interactions with a donor molecule like tetrathiafulvalene having a stronger impact than with an electron-withdrawing molecule like tetracyanoethylene. Although, flaws and edge effects are anticipated to play a significant role, the precise cause of magnetism in the graphene samples cannot be determined at this time. Graphene can acquire ferromagnetic and antiferromagnetic properties from imperfections such as stacking faults. Surface flaws cause magnetic hysteresis in all-inorganic nanoparticles at ambient temperature. Different samples have different magnetisation values and other magnetic characteristics. It is not surprising that HG exhibits the best FM properties if we assume that ferromagnetism is mostly the product of edge effects. This sample has the smallest flakes and hence more edges in addition to having the fewest layers. Intriguingly, the HG Raman spectrum has a larger D band to G band intensity ratio than the EG spectrum. It is noteworthy that preliminary research indicates that the EG-H sample exhibits significantly higher magnetism than HG while likely comprising primarily single-layer graphene [57].

The surface modification of graphene sheets allows for fine-tuning of their magnetic characteristics. For example, when a fully hydrogenated graphene sheet is exposed to an external electric field, the hydrogen atoms on one side can be unloaded while remaining on the other. This results in the formation of a half-hydrogenated graphene sheet, where the unpaired electrons in the unsaturated C sites give rise to magnetic moments coupled through extended p-p interactions. In addition, sp^2-hybridized carbon atoms in the sheet each have an unsaturated dangling bond. This gives the sheet the ability to change its surface in a number of ways.

We demonstrate using first-principles calculations that there are several opportunities for us to modify the surface modification for novel attributes due to the flexible bonding features of carbon atoms. For example, according to our research, the system may switch between being a non-magnetic metallic sheet, a magnetic semiconductor with a tiny indirect band gap, and a non-magnetic semiconductor with a huge direct band gap.

When subjected to hydrogen plasma, a metallic graphene sheet can be fully hydrogenated, creating a wide-gap semiconducting graphene sheet. An electric field can also cause a graphene sheet to become partially hydrogenated, creating a magnetic semiconducting graphene sheet. We explore employing F atom for the surface modification in the half-hydrogenated graphene sheet to fine-tune the characteristics better. We discovered that the band structure for this graphene sheet with the initials F-G-H indicates that it is a straight band gap semiconductor without magnetic characteristics [58]. Researchers from various fields, including magnetic fluids, data storage, biotechnology/biomedicine, catalysis, magnetic resonance imaging, environmental remediation, etc., have recently shown a significant interest in magnetic nanoparticles. Due to the existence of iron cations in two valence states, Fe^{2+} and Fe^{3+}, in the inverse spinel structure, magnetite Fe_3O_4 nanoparticles stand out among the others as having the most intriguing features. Chen and colleagues used chemical precipitation to create a superparamagnetic graphene oxide/Fe_3O_4 nanocomposites, which showed promise for controlled targeted drug release [59]. The superparamagnetic properties of the graphene/Fe_3O_4 nanocomposites open up possibilities for applications in biomedicine, biomaterials separation, and bio diagnostics. Ongoing research is being done in-depth on these applications [63]. It is generally known that modifying the magnetic characteristics of materials by adsorbing or doping foreign atoms is a potential strategy.

First-principles density-functional theory was used to analyse graphene with noble metal (NM = Pt, Ag, and Au) atom adsorption. Given that the Pt atom has an unsaturated electronic d-shell, the Pt-graphene contact is stronger than the interactions between Ag-graphene and Au-graphene (d^9s^1). While the Pt cluster on the graphene does so, the single Pt adatom cannot create a magnetic moment. As a result, the Pt-graphene system eventually became nonmagnetic. The Au-graphene and the Ag-graphene systems, on the other hand, continue to be magnetic because the adsorptive Ag and Au atoms still have magnetic moments and cause magnetisms on the C atoms in the graphene. By adsorbing various NM atoms on the graphene, the magnetic properties of the adatom-graphene system can be changed.

Potential uses for controlling the magnetic characteristics of adatom-graphene systems include electrical and spintronic devices [60]. Significant attention has been generated by reports of high-temperature ferromagnetism in graphene and other materials derived from graphite by many researchers. Defects and edge states, a crucial component of graphene and graphene nanoribbons are responsible for the magnetism in these materials.

There is a lot of curiosity about the occurrence of high-temperature ferromagnetism in materials connected to graphite. It has been established that graphene exposed to radiation exhibits ferromagnetism—further reports of magnetic hysteresis and ferromagnetic-like behaviour in the glassy carbon created by laser ablation. The ferromagnetism found in such nanomaterials is thought to be caused by localised unpaired spins resulting from topological and bonding defects. High-temperature ferromagnetism in microporous carbon results from topological disorder linked to bent graphene sheets. Highly oriented pyrolytic graphite materials have been reported to exhibit room-temperature ferromagnetism. Macroscopic amounts of magnetic graphite (MG) are created by carefully controlling the etching of a closed system using a redox reaction. Due to topographic flaws, MG displays a robust ferromagnetic reaction at room temperature.

Depending on the makeup of the guest species, the adsorption of various guest molecules on graphene results in a reversible low-spin/high-spin magnetic switching phenomenon. The magnetism of nanographene is decreased by water adsorption, reaction with acids, and intercalation with potassium clusters. The interaction with lone pair orbitals and charge transfer with graphene sheets have been theorised as the causes of the reduced magnetism. Magnetism can develop at the edge sites involved in host-guest interactions. The pliable nanographene domains are mechanically compressed by the guest molecules accumulated through physisorption, greatly reducing the space between the nanographene sheets. By aligning the magnetic moments in anti-parallel, such a decrease in the intersheet distance could lower the net magnetic moment. The adsorption of molecules, particularly hydrogen, impacts the ferromagnetism of graphene at room temperature. Graphene-like systems undergo hydrogenation to become semiconducting, and hydrogen adsorption changes these materials' edge configurations and electrical characteristics. Since the creation of tetrahedral carbons can lower the connectivity of the sheets and the energy gap of the localised double bonds and consequently the current ring diamagnetism, hydrogenation of graphene can induce magnetism. Therefore, such structural modifications could result in a rise in magnetic susceptibility (Table 4).

We may state that, taken as a whole, a large number of investigations on few-layer graphenes demonstrate the existence of room-temperature ferromagnetism. Although, imperfections can be significant, the function of edges in these materials appears to be crucial. The magnetic characteristics of graphene are impacted by the adsorption of molecules that act as electron donors and acceptors. The interaction with hydrogen greatly increases few-layer graphenes' magnetism. Another indication of the significance of edge effects and defects is the observation of room-temperature ferromagnetism in graphene analogues of other inorganic materials.

Table 4 Magnetic properties of the hydrogenated graphene samples [61]

Sample name	Wt (%) of hydrogen	M (emu g^{-1}) (FC @ 10 K)		Saturation magnetisation (M_s)	
		3000 Oe	1 T	100 K	300 K
HG		0.017	0.025	0.012	0.01
HGH_1	2	0.034	0.07	0.015	0.012
HGH_2	3	0.044	0.092	0.024	0.02
HGH_3	5	0.048	0.099	0.022	0.016
HGH_2_DH	–	0.038	0.048	0.019	0.015

Significant magnetoresistance in graphene-like materials points to potential device applications [61].

References

1. Lee, C., et al.: Measurement of the elastic properties and intrinsic strength of monolayer graphene. Science **321**(5887), 385–388 (2008)
2. Ovid'Ko, I.A.: Mechanical properties of graphene. Rev. Adv. Mater. Sci. **34**(1), 1–11 (2013)
3. Faber, K.T., Evans, A.G.: Crack deflection processes—I. Theory. Acta Metall. **31**(4), 565–576 (1983)
4. Atif, R., Shyha, I., Inam, F.: Mechanical, thermal, and electrical properties of graphene-epoxy nanocomposites—a review. Polymers **8**(8), 281 (2016)
5. Corcione, C.E., Freuli, F., Maffezzoli, A.: The aspect ratio of epoxy matrix nanocomposites reinforced with graphene stacks. Polym. Eng. Sci. **53**(3), 531–539 (2013)
6. Ramos-Galicia, L., et al.: Improved performance of an epoxy matrix as a result of combining graphene oxide and reduced graphene. Int. J. Polym. Sci. **2013**, (2013)
7. Li, Z., et al.: The role of functional groups on graphene oxide in epoxy nanocomposites. Polymer **54**(21), 5821–5829 (2013)
8. Liu, W., et al.: Simultaneous catalysing and reinforcing effects of imidazole-functionalised Graphene in anhydride-cured epoxies. J. Mater. Chem. **22**(35), 18395–18402 (2012)
9. Yang, H., et al.: Convenient preparation of tunably loaded chemically converted graphene oxide/epoxy resin nanocomposites from graphene oxide sheets through two-phase extraction. J. Mater. Chem. **19**(46), 8856–8860 (2009)
10. Naebe, M., et al.: Mechanical property and structure of covalent functionalised graphene/epoxy nanocomposites. Sci. Rep. **4**(1), 1–7 (2014)
11. Qi, B., et al.: Mechanical and thermal properties of epoxy composites containing graphene oxide and liquid crystalline epoxy. Fibers Polym. **15**(2), 326–333 (2014)
12. Ren, F., et al.: In situ polymerisation of graphene oxide and cyanate ester–epoxy with enhanced mechanical and thermal properties. Appl. Surf. Sci. **316**, 549–557 (2014)
13. Qi, B., et al.: Enhanced thermal and mechanical properties of epoxy composites by mixing thermotropic liquid crystalline epoxy grafted graphene oxide. Express Polym. Lett. **8**(7), (2014)
14. Lu, S., et al.: Epoxy nanocomposites filled with thermotropic liquid crystalline epoxy grafted graphene oxide. RSC Adv. **3**(23), 8915–8923 (2013)
15. Shen, X.-J., et al.: The reinforcing effect of graphene nanosheets on the cryogenic mechanical properties of epoxy resins. Compos. Sci. Technol. **72**(13), 1581–1587 (2012)
16. Bao, C., et al.: In situ preparation of functionalised graphene oxide/epoxy nanocomposites with effective reinforcements. J. Mater. Chem. **21**(35), 13290–13298 (2011)

17. Papageorgiou, D.G., Kinloch, I.A., Young, R.J.: Mechanical properties of graphene and graphene-based nanocomposites. Prog. Mater Sci. **90**, 75–127 (2017)
18. Zhang, P., et al.: Fracture toughness of Graphene. Nat Commun **5**, 3782 (2014)
19. Blees, M.K., et al.: Graphene kirigami. Nature **524**(7564), 204–207 (2015)
20. Tohei, T., et al.: Debye temperature and stiffness of carbon and boron nitride polymorphs from first principles calculations. Phys. Rev. B **73**(6), 064304 (2006)
21. Benedict, L.X., Louie, S.G., Cohen, M.L.: Heat capacity of carbon nanotubes. Solid State Commun. **100**(3), 177–180 (1996)
22. Pop, E., Varshney, V., Roy, A.K.: Thermal properties of graphene: fundamentals and applications. MRS Bull. **37**(12), 1273–1281 (2012)
23. Guo, Z., Zhang, D., Gong, X.-G.: Thermal conductivity of graphene nanoribbons. Appl. Phys. Lett. **95**(16), 163103 (2009)
24. Yan, Z., Nika, D.L., Balandin, A.A.: Thermal properties of graphene and few-layer graphene: applications in electronics. IET Circuits Devices Syst. **9**(1), 4–12 (2015)
25. Kumar, M., et al.: Tuning the thermoelectric material's parameter: a comprehensive review. J. Nanosci. Nanotechnol. **20**(6), 3636–3646 (2020)
26. Kusmartsev, F., et al.: Application of graphene within optoelectronic devices and transistors. In: Applied spectroscopy and the science of nanomaterials, pp. 191–221. Springer (2015)
27. Yang, W., et al.: The physicochemical properties of graphene nanocomposites influence the anticancer effect. J. Oncol. **2019**, (2019)
28. Srivastava, A., et al.: Potential of graphene-based materials to combat COVID-19: properties, perspectives, and prospects. Mater. Today Chem. **18**, 100385 (2020)
29. Yang, G., et al.: Structure of graphene and its disorders: a review. Sci. Technol. Adv. Mater. **19**(1), 613–648 (2018)
30. Zheng, R., et al.: Transport, matransportpecific heat, internal friction, and shear modulus in the charge ordered La0.25Ca0.75MnO3 manganite. J. Appl. Phys. **94**(1), 514–518 (2003)
31. Berry, V.: Impermeability of graphene and its applications. Carbon **62**, 1–10 (2013)
32. Bunch, J.S., et al.: Impermeable atomic membranes from graphene sheets. Nano Lett. **8**(8), 2458–2462 (2008)
33. Sun, P.Z., et al.: Limits on gas impermeability of graphene. Nature **579**(7798), 229–232 (2020)
34. Leenaerts, O., Partoens, B., Peeters, F.: Graphene: a perfect nanoballoon. Appl. Phys. Lett. **93**(19), 193107 (2008)
35. Mallineni, S.S.K., et al.: Influence of dopants on the impermeability of graphene. Nanoscale **9**(18), 6145–6150 (2017)
36. Tsetseris, L., Pantelides, S.T.: Graphene: an impermeable or selectively permeable membrane for atomic species? Carbon **67**, 58–63 (2014)
37. Su, Y., et al.: Impermeable barrier films and protective coatings based on reduced graphene oxide. Nat. Commun. **5**(1), 1–5 (2014)
38. Marques, A.C.C.: Nanobiosensing Platforms with Sers and MIP-Based Technologies. Universidade NOVA de Lisboa (Portugal) (2021)
39. Paul, W., Sharma, C.P.: Blood compatibility and biomedical applications of graphene. Trends Biomater. Artif. Organs **25**(3), 91–94 (2011)
40. Marzana, M., et al.: Nanocarbon for bioelectronics and biosensing. In: Nanomaterials for biocatalysis, pp. 689–714. Elsevier (2022)
41. Byun, J.: Emerging frontiers of graphene in biomedicine. J. Microbiol. Biotechnol. **25**(2), 145–151 (2015)
42. Priyadarsini, S., et al.: Graphene and graphene oxide as nanomaterials for medicine and biology application. J. Nanostructure Chem. **8**(2), 123–137 (2018)
43. Li, D., et al.: When biomolecules meet graphene: from molecular level interactions to material design and applications. Nanoscale **8**(47), 19491–19509 (2016)
44. Yao, J., et al.: Recent advances in graphene-based nanomaterials: properties, toxicity and applications in chemistry, biology and medicine. Microchim. Acta **186**(6), 1–25 (2019)
45. Kumar, P., et al.: Antibacterial properties of graphene-based nanomaterials. Nanomaterials **9**(5), 737 (2019)

46. Catania, F., et al.: A review on recent advancements of graphene and graphene-related materials in biological applications. Appl. Sci. **11**(2), 614 (2021)
47. Fuchs, J.-N., Goerbig, M.O.: Introduction to the physical properties of graphene. Lect. Notes **10**, 11–12 (2008)
48. Huang, X., et al.: Graphene-based materials: synthesis, characterisation, properties, and applications. Small **7**(14), 1876–1902 (2011)
49. Avouris, P., Dimitrakopoulos, C.: Graphene: synthesis and applications. Mater. Today **15**(3), 86–97 (2012)
50. Abergel, D., et al.: Properties of graphene: a theoretical perspective. Adv. Phys. **59**(4), 261–482 (2010)
51. Kumar, R., et al.: Laser processing of graphene and related materials for energy storage: state of the art and future prospects. Prog. Energy Combust. Sci. 100981 (2022)
52. Ioniţă, M., et al.: Graphene and functionalised graphene: extraordinary prospects for nanobiocomposite materials. Compos. B Eng. **121**, 34–57 (2017)
53. Wei, W., Qu, X.: Extraordinary physical properties of functionalised graphene. Small **8**(14), 2138–2151 (2012)
54. Stolyarova, E., et al.: Observation of graphene bubbles and effective mass transport under graphene films. Nano Lett. **9**(1), 332–337 (2009)
55. Ghaffarkhah, A., et al.: Synthesis, applications, and prospects of graphene quantum dots: a comprehensive review. Small **18**(2), 2102683 (2022)
56. Tang, N., et al.: Magnetic properties of graphene. In: Spintronic 2D materials, pp. 137–161. Elsevier (2020)
57. Matte, H.R., Subrahmanyam, K.S., Rao, C.N.: Novel magnetic properties of Graphene: presence of both ferromagnetic and antiferromagnetic features and other aspects. J. Phys. Chem. C **113**(23), 9982–9985 (2009)
58. Zhou, J., et al.: Tuning electronic and magnetic properties of graphene by surface modification. Appl. Phys. Lett. **95**(10), 103108 (2009)
59. Jiang, C., et al.: Recent advances in graphene-family nanomaterials for effective drug delivery and phototherapy. Expert Opin. Drug Deliv. **18**(1), 119–138 (2021)
60. Tang, Y., et al., Noble metals induced magnetic properties of Graphene. J. Magn. Magn. Mater. **323**(20), 2441–2447 (2011)
61. Rao, C., et al.: Unusual magnetic properties of graphene and related materials. Chem. Sci. **3**(1), 45–52 (2012)

Fabrication Routes of Graphene

Then Mun Yip and Goh Boon Tong

Abstract This chapter describes various state-of-the-art fabrication routes for high-quality graphene included: chemical vapour deposition (CVD), mechanical exfoliation, chemical exfoliation, electrochemical exfoliation, arc discharge, epitaxial growth, and pyrolysis. CVD is a widely used technique for growing high-quality graphene films on metal catalyst substrates, and copper foil has shown promising results. Mechanical exfoliation involves peeling graphite flakes from highly oriented pyrolytic carbon (HOPG) platelets using Scotch tape, resulting in single-layer graphene. Chemical exfoliation has two methods: solution-assisted and low-temperature chemical exfoliation. Electrochemical exfoliation involves the intercalation and exfoliation of graphite into graphene nanosheets through electrolyte solutions. Arc discharge is a plasma deposition technique for synthesizing high-quality graphene sheets using alternating current arc-discharge processes. Epitaxial growth involves growing single-layer or multilayer graphene on a SiC substrate using high-temperature sublimation growth. Pyrolysis is a 6-step process of poly(methyl methacrylate) composite that results in carbon derivatives that dissolve in the Ni catalyst surface, resulting in the epitaxial growth of graphene. Each method has its unique features, advantages, and disadvantages, making them suitable for different applications. For example, mechanical exfoliation remains one of the most reliable ways of producing high-quality graphene and has led to the discovery of graphene's extraordinary physical properties. Chemical exfoliation can produce graphene on a large scale, and electrochemical exfoliation is effective in creating biocompatible and fluorescent carbon nanomaterials for biological labelling and imaging. CVD as well as epitaxial growth can produce high-quality graphene films, and pyrolysis produces graphene with a high degree of graphitization. The choice of the appropriate technique is crucial for specific applications.

Keywords Graphene · CVD · Exfoliation · Arc discharge · Pyrolysis

---

T. M. Yip · G. B. Tong (✉)
Low Dimensional Materials Research Centre, Department of Physics, Faculty of Science, University of Malaya, 50603 Kuala Lumpur, Malaysia
e-mail: gohbt@um.edu.my

© The Author(s), under exclusive license to Springer Nature Singapore Pte Ltd. 2023

R. T. Subramaniam et al. (eds.), *Graphene*, Engineering Materials,
https://doi.org/10.1007/978-981-99-1206-3_4

1 Chemical Vapour Deposition (CVD)

The CVD technique is a procedure in which chemical gases or vapours react on the substrate surface to deposit coatings or nanostructures [11]. Uniform deposition, high degree of control, good quality, and cheap cost are all benefits of the CVD process. It has been extensively employed in the creation of one of the most important carbon materials, which is carbon nanotubes (CNTs). In the 1960s, the first effort to create graphitic layers using the CVD process was made [69]. However, a lack of an appropriate transferring method and investigation approach at the time resulted in a restricted knowledge of as-grown graphitic layers. Hong et al. devised a straightforward way to transfer and measure graphene on pattern nickel (Ni) sheets using polydimethylsiloxane (PDMS) films as intermediate adhesive layers in 2009, which coincided with the rapid progress of graphene CVD growth [35].

However, since the graphene precipitated on Ni sheets had a multilayer structure, its physical characteristics differed significantly from those of monolayer graphene. After that, the copper (Cu) foil was discovered to be the most favourable substrate for fabricating monolayer graphene with a self-limiting growth function due to the ultralow dissolution of carbon, paving an effective route for the preparation of the monolayer graphene film using the CVD method, which attracted a lot of attention [41].

The growth process is substantially determined by the metal catalysts utilzsed. Initial CVD development of graphene on Ni foils was reported since it was widely known that carbonization over Ni metal creates graphite flakes [61]. C atoms decomposed from CH_4 molecules are readily dissolved into metal films with high carbon solubility, such as Ni and Co, and precipitate as graphene during the final cooling phase. Due to the difficulty in controlling the quantity of dissolved C atoms, these metals produce nonuniform graphene sheets with inhomogeneous layer numbers.

Ruoff's team revealed in 2009 that Cu catalyst may create a homogenous single-layer graphene (single-layer fills 95% of total area) [41]. Owing to the poor carbon solubility, the dissolution of C atoms into Cu metal is inhibited, and graphitization occurs solely on the Cu surface, resulting in the preferred formation of a single-layer graphene [48]. The CH_4 breakdown stops after the Cu is covered by graphene; this is referred to as a "self-limiting process." Cu metal, namely Cu foil, is currently commonly employed as a graphene growth catalyst because a single-layer graphene may be generated across a large reaction window. In theory, large-area graphene may be produced by scaling up both the catalyst substrate and the CVD furnace. This was shown by Hong and Ahn's groups, who were able to synthesize a 30-inch graphene sheet atop Cu foil and report roll-to-roll transfer to a polymer film [2]. They also proved the potential of CVD graphene by manufacturing touch screens from a graphene sheet that had been transferred.

Muller's group used a dark-field transmission electron microscope (DF-TEM) to analyse the domain structure of CVD graphene produced on polycrystalline Cu foil in 2011. They discovered that CVD graphene is polycrystalline, with tiny domains ranging in size from 500 nm to few μm that are randomly rotated [27]. Furthermore, mechanical tears are theorised to start at these domain boundaries [34]. Since graphene synthesised by CVD is a patchwork of tiny graphene domains, even if big-area graphene can be grown using huge Cu foil, optimal physical attributes cannot be anticipated from such polycrystalline graphene. For high-performance electronics and other applications, CVD growth of single-crystalline graphene, lack of boundaries is needed. Metal catalysts with polycrystalline or amorphous structures are one of the key reasons for the polycrystallinity of CVD graphene.

Single-crystalline metal catalysts are projected to outperform polycrystalline metal foils or films for high-quality graphene formation via CVD. However, due to the high cost and restricted size of single-crystalline substrates like Cu(111) and Ni(111), it is not viable to create graphene on these substrates, which need a transfer procedure that involves chemical metal etching. A novel technique for epitaxially depositing crystalline metal films on single-crystalline oxide substrates like sapphire (α-Al_2O_3) and MgO is devised [24]. These oxide substrates are substantially less expensive than Cu(111) and Ni(111) substrates, and due to the growth of the GaN-based light-emitting diode (LED) sector, big sapphire wafers up to 6 inches are now accessible at a reasonable price.

A single-layer graphene with a large area was produced on the Cu foil by CVD of CH_4 at up to 1000 °C [41]. The films were mostly single-layer, which was primarily due to Cu's poor carbon solubility (approximately 0.03%) compared to Ni's high solubility (about 1.1%). The grain boundaries and steps of Cu contribute to double- and triple-layered graphene flakes with wrinkles that were observed. TEM pictures of folded edges of the single-layer graphene (1L) and two-layered graphene (2L) are shown in the insets (2L). Wrinkles caused by the difference in Cu and graphene thermal expansion coefficients also cross Cu grain boundaries, showing that the graphene layer is continuous. The graphene grains exhibit no obvious epitaxial connection to the Cu substrate when grown on a polycrystalline Cu substrate [85]. Transparent conductive films with a large area (23×20 cm^2) of graphene layers were created using surface wave plasma CVD of CH_4 mixed with Ar and/or H_2 over a Cu or Al foil substrate at low temperatures such as 300–400 °C [33]. CVD of ethanol (C_2H_5OH) vapour at 700–800 °C grew a patterned graphene sheet atop a patterned Cu layer on a SiO_2/Si substrate, which was used as the electrodes for field-effect transistors [13]. CVD of CH_4 was used to generate graphene on the (100) face of a high-purity Cu single crystal, revealing defect-free graphene formation [60]. The defect density and homogeneity of graphene films produced on Cu foil from C_2H_5OH were found to be lower than those formed from pentane (C_5H_{12}) [86]. C_2H_5OH-derived films have a sheet resistivity of 2700 U/sq and a Hall mobility of 110 cm^2/V, while C_5H_{12}-derived films have 5000 Ω/sq and 65 cm^2/V, respectively.

Co films were produced on sapphire and SiO_2 substrates by CVD at 1000 °C [19]. The 200 nm-thick Co film sputtered on sapphire at room temperature was discovered to be polycrystalline. This polycrystalline Co film was unable to endure the 1000 °C high-temperature CVD procedure, resulting in many pits and holes on the metal surface. However, high-temperature Co sputtering (500 °C) and post-annealing in H_2 dramatically increased crystallinity and stability during CVD. The thin Co film produced on a Si wafer with a 300 nm amorphous oxide layer, on the other hand, was polycrystalline and unstable throughout the CVD process.

Cu foil put over an oxide substrate with a gap of roughly 15 μm allowed for a rapid growth rate of massive single-crystal graphene films by CVD of CH_4 at 1000 °C [81]. In contrast to rates of less than 0.4 μm/s (0.03–0.36 μm/s), it reached as high as 60 μm/s [78]. The considerable rise in CH_4 dissociation to CH_3 radicals through the interaction with oxygen on the Cu(100) surface explained the acceleration of graphene formation by oxygen. Through an optimised gap, oxygen was delivered from the oxide substrate [81]. It was feasible to generate millimetre-sized single-crystalline graphene using sequential oxygen passivation during CVD nucleation and growth. The growth rate was about 100 μm/min [7]. The first O_2 passivation of about 0.1 standard cubic centimetres per minute (sccm) was performed in a vacuum of about 1 Pa at 1040 °C, where Cu foil was annealed, followed by the introduction of CH_4/H_2 with 1/300 for nucleation, and then the second O_2 passivation followed by CH_4/H_2 with 1/5 to grow on the graphene layer. The use of O_2 passivation before both the nucleation and growth phases resulted in the creation of large graphene flakes as big as 2 mm, with fewer flakes, as seen in Fig. 1a. Without O_2 during either nucleation or growth in Fig. 1b and Fig. 1c respectively, graphene flakes have a substantially greater nucleation density but are much smaller, around 2 and 20 mm in diameter. This accelerating impact of O_2 was expected by a considerable decrease in the adsorption energy of carbon atoms on the oxygen-covered Cu surface, according to density functional theory (DFT) calculations. By using alcohols as precursors or adding water vapour to the precursor gases during the catalytic CVD synthesis of carbon nanotube (CNT) arrays (or forests), the growth rate of CNT arrays (or forests) was significantly accelerated [21].

CVD was used to deposit graphene sheets on a Cu foil at 1000 °C with CO_2 mixed in H_2 gas [45]. CO_2 gas was activated at 200 °C using a Ni/Al_2O_3 catalyst before being mixed with 200 sccm into the H_2 flow. The graphene film's shape and lateral size varied as the CO_2 flow rate increased, from a hexagonal 6 μm film for a 5-sccm rate of CO_2 to a circular 1 μm film for a 30 sccm rate of CO_2. The maximal mobility

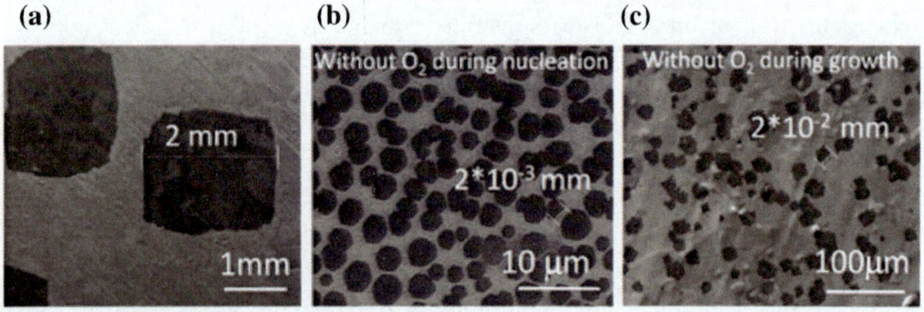

Fig. 1 SEM images of graphene grown on Cu foil: **a** with O_2 during both nucleation and growth **b** without O_2 during nucleation and **c** without O_2 during growth [7]

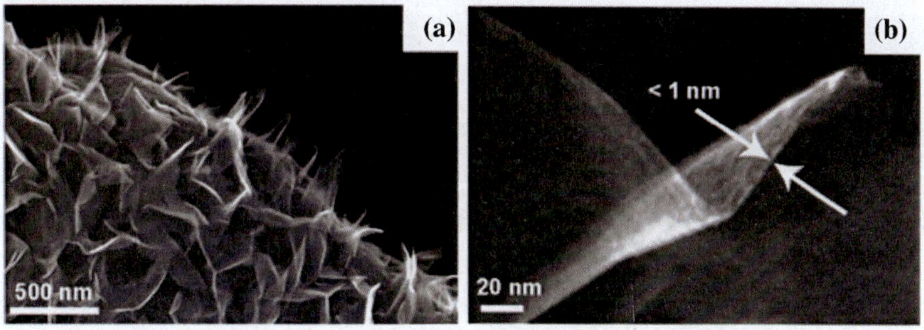

Fig. 2 Graphene nanoflakes grown vertically to the substrate [88]

of holes and electrons in the synthesised graphene was established via field-effect transistors to be 3010 cm^2/V and 750 cm^2/V, respectively.

Radio-frequency (RF) plasma-enhanced CVD of CH_4 diluted in H_2 with a substrate temperature Ts of 600–900 °C yielded graphene nanoflakes grew virtually vertically to several substrates, including Si, Ni, and graphite [74]. The nanoflakes are displayed in SEM images in Fig. 2. A 1–1.5-nm-thick carbon layer was first formed parallel to the substrate surface, after which the leading edge of the top layer coiled up and became vertical [88]. The growth rate of nanoflakes increased as the substrate temperature increased, and the vertical development of nanoflakes was disrupted. On these vertically produced graphene nanoflakes, a low turn-on field of 4.7 V/μm was reported for electron field emission [74]. A 120 Hz current was efficiently filtered by an electric double-layer capacitor with these nanoflakes produced directly on metal current collectors [52].

The properties of graphene generated with the third placement approach were investigated. Figures 3a and b show optical microscopy and SEM pictures of graphene which was deposited onto SiO_2/Si, respectively. A few polycrystalline graphene grains with an average size of 1.5 μm (area A) and a large number of 260-nm nanocrystalline grains make up the as-grown graphene (area B). The Raman spectroscopy results for regions A and B in Fig. 3b, which both belong to graphene, are shown in Fig. 3c. The Raman spectrum's extremely faint D peak in region A suggested that there were very few flaws in this location. The I_{2D}/I_G ratio of 1.17 indicated that there were just a few layers. The Raman spectrum's high-intensity D peak in region B indicated that there were many edges and flaws in the area, which were created by nanometer-sized graphene grains. The AFM surface morphology of the graphene layer is shown in Fig. 3d. The hexagonal graphene grains were micron-sized and surrounded by spherical nanocrystalline graphene. As shown in Fig. 3e, the thickness of the white line was around 1.5 nm, suggesting that there were 1–3 graphene layers. In summary, after 60 s of growth at 850 °C, graphene with two grain sizes was formed. Both micron-scale polycrystalline and nanocrystalline graphene forms were presented in the as-grown graphene.

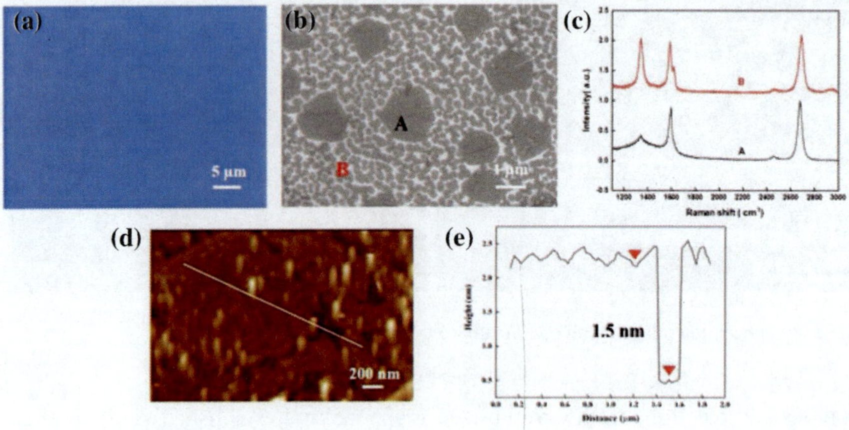

Fig. 3 Graphene grown onto SiO_2/Si: **a** optical microscope image **b** SEM image **c** Raman spectra at points A and B in (**b**) **d** AFM image **e** The white line's height diagram in (**d**) [39]

2 Mechanical Exfoliation

Geim's research group produced the first single-layer graphene sample via mechanical exfoliation in 2004 [30]. It was a straightforward procedure involving the use of common Scotch tape as the primary experimental tool. In their work, they used highly orientated pyrolytic carbon (HOPG) platelets as the raw material. On the top of the HOPG platelets, several 5-m-deep thick mesas in diameters ranging from 20 to 2 mm were created using dry oxygen plasma etching. After that, the structured surface was pressed against a wet photoresist layer on a glass substrate. After baking, the researchers split the HOPG sample from the substrate, leaving only mesas that were tightly connected to the photoresist layer. Using Scotch tape, they began peeling graphite flakes from the mesas. The graphite that remained on the photoresist became thinner and thinner after repeated peeling. Finally, dissolution of the photoresist produced ultrathin graphite flakes linked to the photoresist in acetone. The solution was dipped into a Si wafer with a 300-nm thick SiO_2 layer on top, and some flakes were caught by the Si wafer. After removing relatively thick flakes from the Si wafer with ultrasonic, ultrathin flakes strongly adhered to Si were obtained for characterization. Under optical microscopes, graphite flakes of various thicknesses on a Si substrate with a SiO_2 layer on the top appear to have varied hues. As a result, it can distinguish thin exfoliated graphite flakes from thick ones by observing them with the naked eye, even though single-layer graphene is nearly invisible in optics. Using atomic force microscopy, researchers were able to identify the single-layer graphene with a thickness of 1 nm amid a large number of graphite flakes. The single-layer sample was then characterised, revealing nearly miraculous and astonishing physical features of graphene, a unique carbon substance that has since gained worldwide acclaim.

Teng et al. also demonstrated mechanical exfoliation of graphite for the manufacture of graphene using zirconium oxide (ZrO_2) balls with diameters of 2 mm and 0.2 mm, respectively, for big and tiny balls [68]. Graphite is ball milled through ZrO_2 balls by crashing graphite between huge ZrO_2 balls. Small ZrO_2 balls, on the other hand, crush shorter graphite particles and shear them into many thin graphene sheets, as depicted in Fig. 4. As a result, this approach produces exfoliated graphene by varying the surface area of ZrO_2 balls, resulting in a huge number of exfoliated graphene.

Apart from that, Soldano and his collaborators have also attempted to demonstrate this technique and Fig. 5 shows the optical results of their research. The apparent contrast of the graphene monolayer on a SiO_2/Si substrate (with an oxide thickness of either 300 or 90 nm) was maximum at roughly 12% at 550 nm, indicating that the substrate choice is significant. This was explained using a Fabry–Perot multilayer cavity in which the optical path contributed by graphene to the SiO_2/Si system's interference reached greatest for particular oxide thicknesses [4]. Thicker graphite flakes placed over a 300 nm SiO_2 layer appeared yellow to bluish as the thickness decreased (Fig. 5a), whereas few- or one-layer graphene seemed deeper to lighter purple colours (Fig. 5b) [62]. It is worth noting that the tape method can leave glue residues on the substrate surface, limiting carrier mobility [8]. The mobility of exfoliated graphene samples on SiO_2 substrates ranges from 5000 to 30,000 $cm^2\ V^{-1}\ s^{-1}$.

Then, mechanical exfoliation has also been reported to produce a graphene nanosheet with a thickness of only one atom [51]. At an acceleration voltage of only 80 kV, 1-A resolution was attained by combining aberration correction with a monochromator. As a result, each carbon atom in the area of vision was identified and resolved. A highly crystalline lattice was discovered, along with a few point flaws. Stone-Wales flaws were seen in situ during their creation and annealing. Figure 6a depicts an optical micrograph of a large graphene sample on the support grid, with the 1-μm grid perforation holes clearly visible, whereas Fig. 6b depicts a low-magnification TEM picture of a single-layer graphene nanosheet, which covers multiple holes. Figure 6c shows a high-resolution TEM picture of one of the suspended nanosheet sections acquired by zooming in. The image was a single unaltered CCD exposure in which the intensity profile was a direct representation of

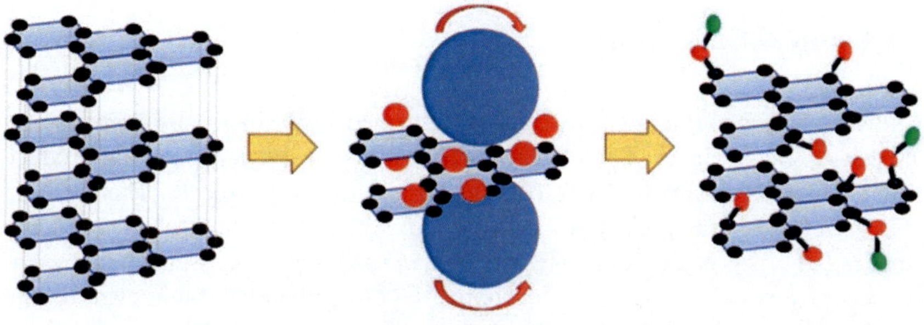

Fig. 4 Mechanism for exfoliation of graphite using a mixture of big and tiny ZrO_2 ball milling [5]

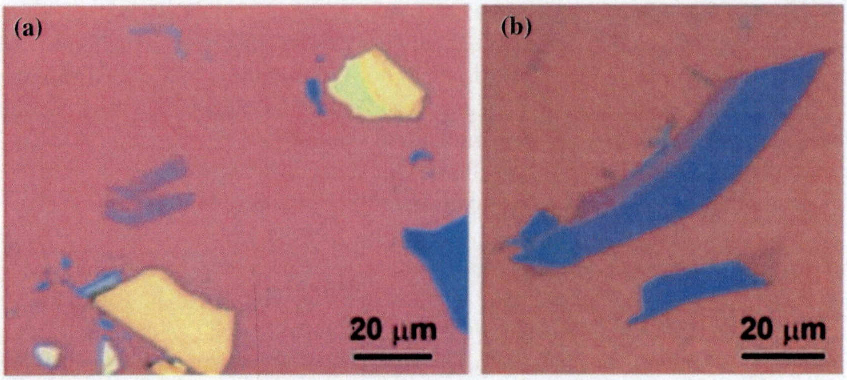

Fig. 5 Optical pictures of **a** thin graphite, **b** few layer graphene, and single-layer graphene (lighter purple contrast) on a 300 nm SiO_2 layer micromechanically exfoliated graphene. Thicker samples are indicated by a yellow tint, whilst thinner samples are indicated by a bluish and lighter contrast [62]

graphene's carbon atomic structure, with the carbon atoms in white. The vast area depicted in Fig. 6c had complete structural integrity. Figure 6d shows a direct picture of a single-layer graphene nanosheet.

Mechanical cleavage is a good method for obtaining excellent graphene samples for fundamental study because it does not affect the atomic structure of graphene. Mechanical exfoliation, on the other hand, has a very poor efficiency. Although, various attempts have been made to replace the manual exfoliation technique pioneered by Geim and his collaborators with a machine-based process [55], the efficiency has not increased significantly, and the yield of high-quality single-layer graphene remains low. Although, the samples of graphene obtained could be beneficial for fundamental investigations to define their chemistries and properties, this technology is neither scalable nor capable of bulk production. As a result, more efficient graphene preparation processes must be developed in order for it to be used on a large scale.

3 Chemical Exfoliation

As illustrated in Fig. 7, Stankovich et al. [63] showed a solution-assisted method for synthesising a single layer of graphene. Graphite is first chemically treated to produce a water-dispersible graphitic oxide (GO) intermediate in this process. The resulting GO is a stacked layer of puckered sheets that exfoliate fully when mechanical energy is applied [31]. This is due to the intensity of interactions between water and oxygen-containing (epoxide and hydroxyl) functionalities brought into the basal plane during oxidation. Water may easily intercalate between the sheets and scatter them individually because of their hydrophilicity. Despite the fact that GO is nonconducting,

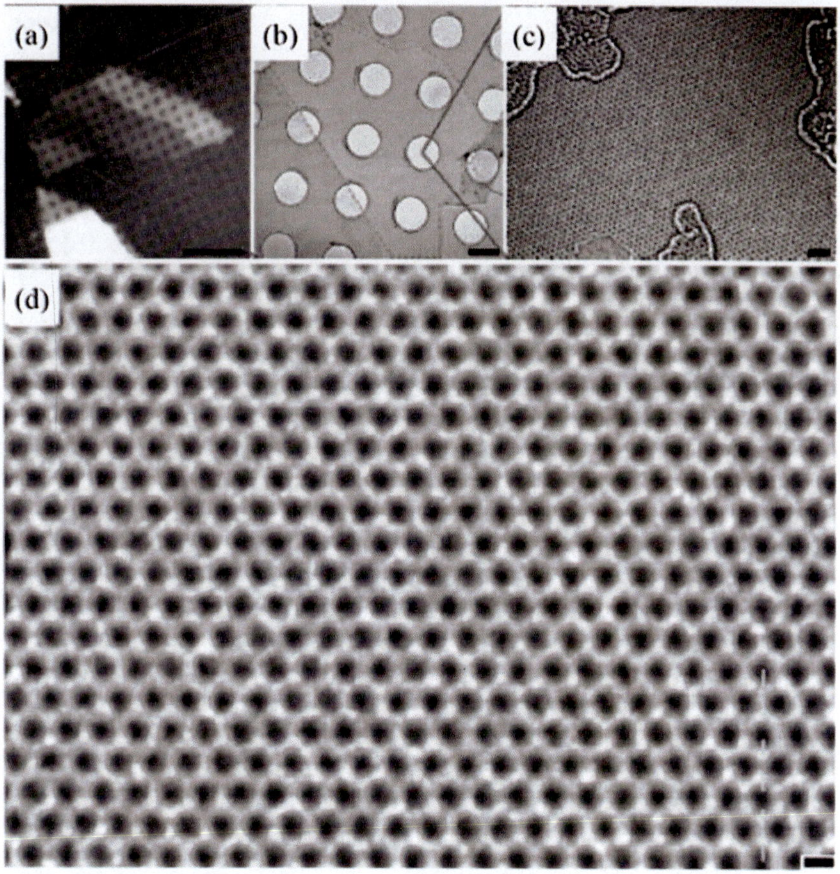

Fig. 6 **a** Optical micrograph and **b** low-magnification TEM image of the graphene sheets on perforated carbon film. A single-layer region is outlined by the red dashed line. **c** Unfiltered CCD exposure (1 s) of a single-layer graphene nanosheet. The structures near the edge of the image are adsorbates and a hole (formed after prolonged irradiation) is observed near the lower-edge left. Scale bars are 10 μm (**a**), 1 μm (**b**) and 1 nm (**c**). **d** Direct image of a single-layer graphene nanosheet, with carbon atoms being white [51]

the graphitic network may be significantly reconstituted by either thermal annealing or chemical reducing agents, both of which have been investigated. According to Stankovich et al. [63], hydrazine hydrate is the best reagent for preventing oxidation by forming and removing epoxide complexes, which requires adding hydrazine directly to aqueous GO dispersions. The initial aqueous reduction of GO, on the other hand, resulted in the elimination of oxygen groups, making the reduced sheets less hydrophilic and more easily aggregated in solution. Even in the deoxygenated sheets, increasing the pH during the reduction produces charge stabilised colloidal dispersions. Tung et al. [71] enhanced the reduction stage by creating dispersions in anhydrous hydrazine directly. Since hydrazine is both very poisonous and possibly

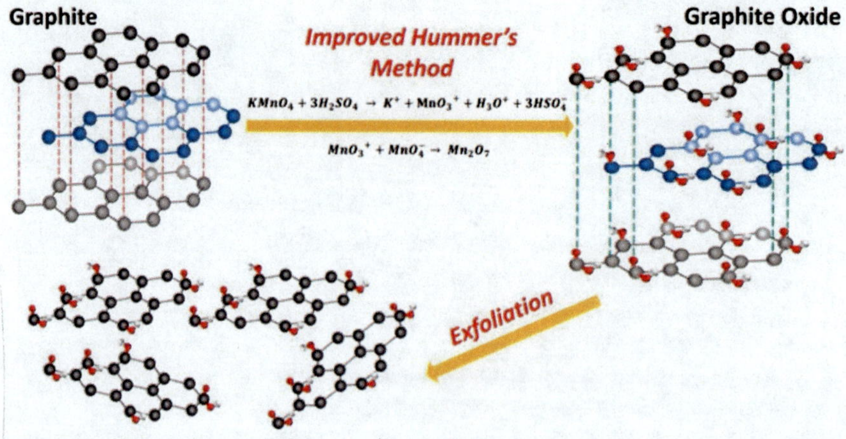

Fig. 7 Conversion process from graphite to chemically derived graphene [20]

explosive, it must be used with extreme caution. The GO method's cheap cost and huge scalability are the most interesting features. The initial material is a simple graphite, and the method may be readily scaled up to generate greater quantities of chemically produced graphene dispersed in a liquid [38].

Lv et al. [47] achieved chemical exfoliation at a very low temperature, much below the predicted critical exfoliation temperature, while working in a high vacuum atmosphere. The graphite oxide exfoliation temperature was as low as 200 °C. The high vacuum environment helped to speed the growth of graphene layers by applying an outward pulling force, which was largely responsible for the effective exfoliation, as shown in Fig. 8. The first phase of a chemical exfoliation of graphene entails employing a modified Hummers process to produce completely oxidised graphite, and the second step involves the release and stability of individual layers under high vacuum [28]. The reduction or elimination of oxygen takes place in a quartz tube under a high vacuum (1 Pa). The graphite oxide was retained at 200 °C for 5 h to remove the superabundant functional groups, and a high vacuum (below 1 Pa) was maintained during the heat treatment. The experiment is then repeated at various temperatures of 300 and 400 °C. G-200, G-300, G-400, and G-HT (produced at 1000 °C using a commonly used high-temperature exfoliation process) were the labels assigned to the graphene sample that was created.

The thermal analysis was carried out to investigate the heating-induced structural changes of graphite oxides, and Fig. 9a shows relatively basic thermogravimetric differential scanning calorimetry (TGDSC) profiles. Within a very limited temperature range (150–250 °C), these profiles are characterised by a strong exothermic DSC signal and a corresponding abrupt mass-loss peak in the TG branch. These findings suggest that in this restricted temperature range, the majority of oxygen-containing functional groups bound to graphene planes are eliminated. Although such a low temperature may cause oxygen to escape from planar graphene sheets,

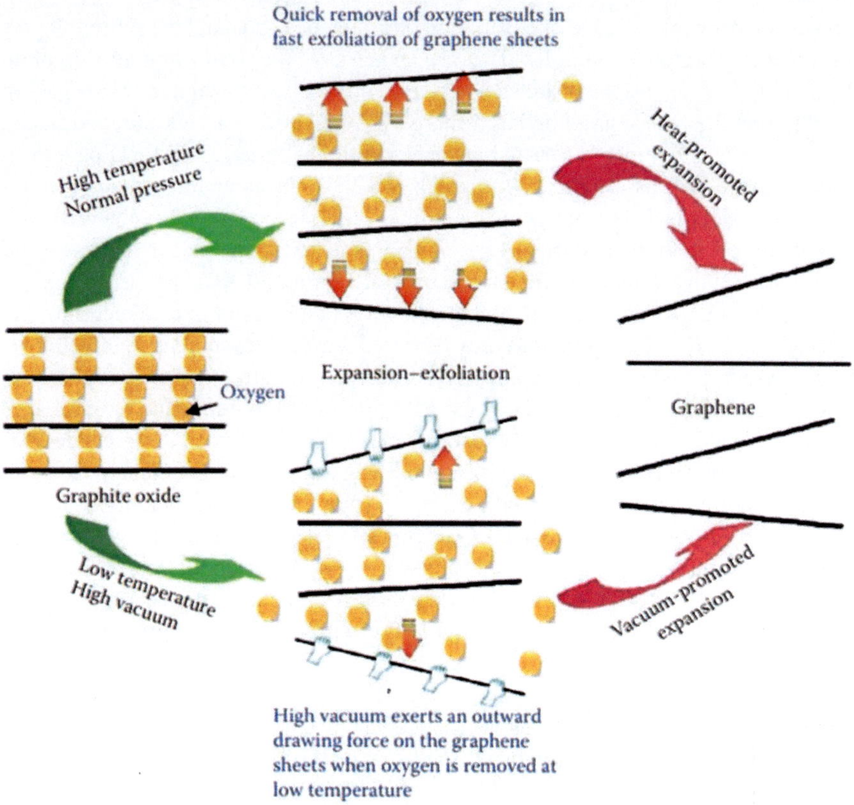

Fig. 8 Mechanism for chemical exfoliation of graphene [47]

it is insufficient to completely expand the graphene layers under atmospheric pressure. In most circumstances, a preheated high temperature environment is required for rapid graphene sheet expansion. If an inner stress generated by the removal of oxygen can be reinforced at the decomposition temperature (150–250 °C) of oxygen-containing groups, a fast expansion exfoliation of graphene layers can be achieved, and the few-layered graphene can be produced at such a low temperature. A vacuum environment (1 Pa) was used to accomplish a quick exfoliation of graphene layers and to stabilise the individual sheets, which led to the discovery of this technique. The high vacuum that pairs with the escape of oxygen at a low temperature such as 200 °C produces an outward pulling force on the growing graphene layers, assisting in the acceleration of graphene layer expansion and resulting in successful graphene layer exfoliation, as exemplified by Fig. 8. The few-layered graphene has been generated in gram scale using this method, and the resulting sample is designated as G-200, which corresponds to the thermal treatment temperature. Thermally processed at 300 and 400 °C under high vacuum, respectively, G-300 and G-400 are made as references. X-ray diffraction (XRD) and microscopic observations were used to determine the degree of exfoliation of graphene oxide and the morphologies of the resulting graphene samples. The XRD patterns of the source graphite, graphite oxide, and resultant graphene are shown in Fig. 9b. With a typical interlayer spacing of 0.335 nm, the parent graphite has a sharp (002) peak at 31.1° (Co-target utilised, peak is upshifted compared to usual instances with Cu target). The XRD pattern of graphite oxide reveals a typical (002) peak at 12.3°, corresponding to a 0.776 nm interlayer spacing. This implies that graphite has been completely changed into graphite oxide, with the majority of the oxygen attached to the planar surface of the graphite following oxidation. The strong peak at 12° on the XRD patterns vanishes after the low-temperature heat treatment, showing that oxygen intercalated into the interlayer spacings of graphite is mainly removed during the vacuum-promoted low-temperature expansion. When compared to parent graphite oxide (Fig. 9c), some typical scanning electron microscopy (SEM) pictures of low-temperature exfoliated samples (Fig. 9d and e) demonstrate that these layers are exfoliated to a great degree. Due to its poor electric conductivity, a good SEM picture of graphene oxide can only be captured when a tiny gold coating is sprayed onto the sample's surface. High resolution imaging of the low temperature exfoliated samples may be readily obtained without any pre-treatments, indicating that the materials have improved conductivity. G-200-HT (heat treating G-200 in a preheated furnace at 1000 °C) and G-HT (produced by a commonly used high-temperature exfoliation process at 1000 °C) are shown in Fig. 9f and g. In terms of morphological observations and adsorption measurements, there is no discernible difference between G-200-HT and G-HT and G-200.

After thoroughly sonicating these sheets with the use of surfactant (sodium dodecyl sulphate, $NaC_{12}H_{25}SO_4$, abbreviated as SDS), the thin-layered graphene was further characterized using AFM, and it was discovered that over 60% of the observed individual sheets had a thickness of less than 1.2 nm. The existence of SDS around the graphene sheets suggests that the majority of these sheets are single-layered [26]. Figure 9h and i show a typical AFM image in which three graphene sheets (circled part) are identified as being roughly overlayered with each other and the corresponding contour profile, which clearly shows that the one-layer part has a depth of 0.78 nm, the two-layered part has a depth of 1.55 nm, and the three-layered part has a depth of 2.36 nm.

4 Electrochemical Exfoliation

Studies on the intercalation of pure sulphuric acid in graphite particulates, electrochemical intercalation of F ions into graphite in aqueous and anhydrous HF media conversion of a $NiCl_2$-based graphite intercalated compounds (GICs) into a more useful $Ni(OH)_2$-based GIC by electrochemical polarisation in alkaline media and electrochemically induced intercalant exchange led to the discovery of electrochemical methods [3]. For electrochemical exfoliation, highly oriented pyrolytic graphite

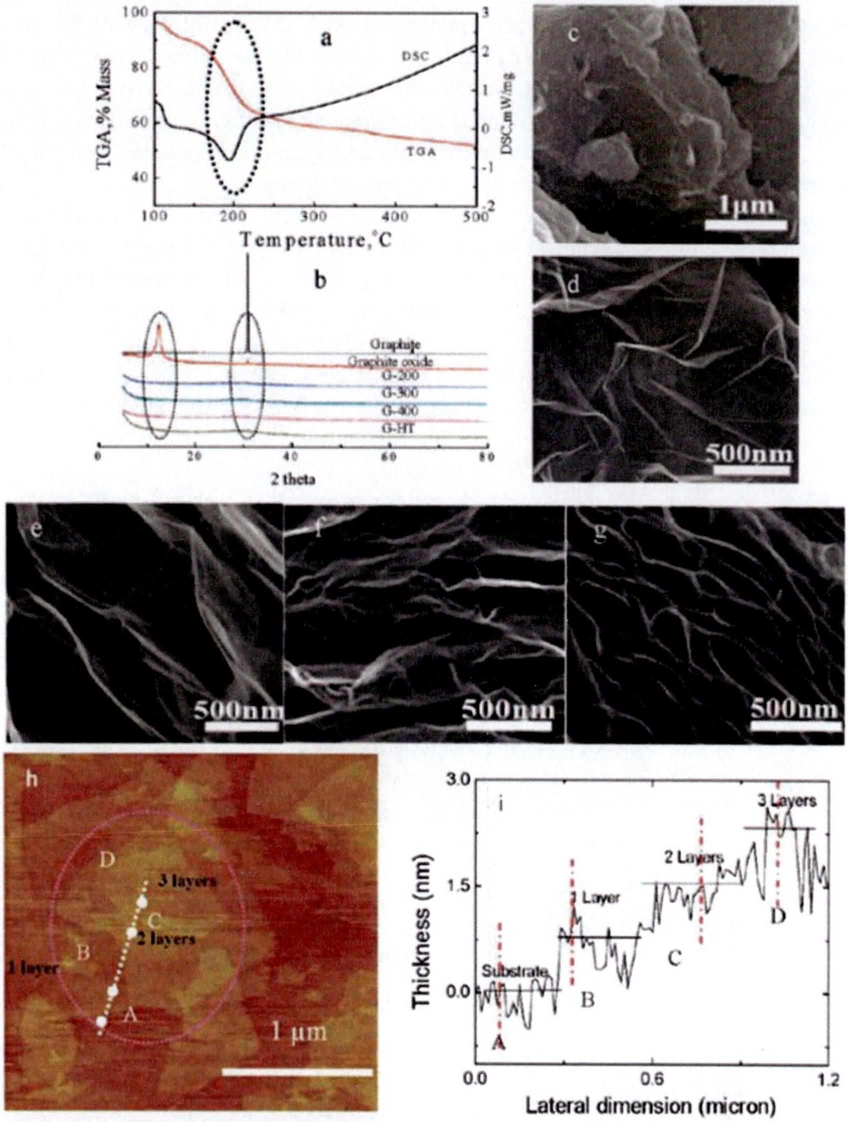

Fig. 9 Characterization of a graphite oxide and the obtained graphene: **a** TG-DSC curves of a graphite oxide; **b** XRD patterns of graphite, graphite oxide and exfoliation-resulting graphene; **c–g** SEM images of parent graphite oxide, G-200, G-400, G-200-HT and G-HT [47] **h** AFM ichnography **i** cross-section contour of G-200 dispersed with the aid of a surfactant (SDS) [47]

(HOPG) and vitreous carbon electrode materials have also been utilised [22]. Electrochemical exfoliation methods use cathodic or anodic potentials, currents, or voltage to exfoliate electrolytes in aqueous (acidic or other media) or nonaqueous solutions. The yield of graphene nanosheets produced determines the scaling-up of electrochemical exfoliation [75]. The ideal method should be capable of producing monolayer graphene nanosheets with high conductivity and a big size.

The following components make up an electrochemical intercalation and exfoliation of graphite into graphene nanosheets experimental setup: (i) a graphite working electrode (WE), (ii) a standard reference electrode or a quasi-reference electrode that varies depending on the electrolyte used, (iii) a counter electrode that is used in a three-electrode arrangement, (iv) an electrolyte solution that can be aqueous (e.g., acidic or surfactant) or non-aqueous (e.g., organic or ionic liquid. A schematic diagram of an electrochemical exfoliation setup for a two-electrode undivided electrochemical cell is shown in Fig. 10. The distinction between two-electrode and three-electrode electrolytic cells is shown in Fig. 11.

Figure 12 shows the three-step process for electrochemical exfoliation. Stage I is when before apparent evidence of exfoliation could be seen, there existed an induction stage. The electrolyte solution's hue went from colourless to yellow, then dark brown. Hydroxyl and oxygen radicals were created by anodic oxidation of water. The dissolution of fluorescent carbon nanocrystals from the anode was caused by the hydroxylation or oxidation of graphite by these radicals, which occurred mostly

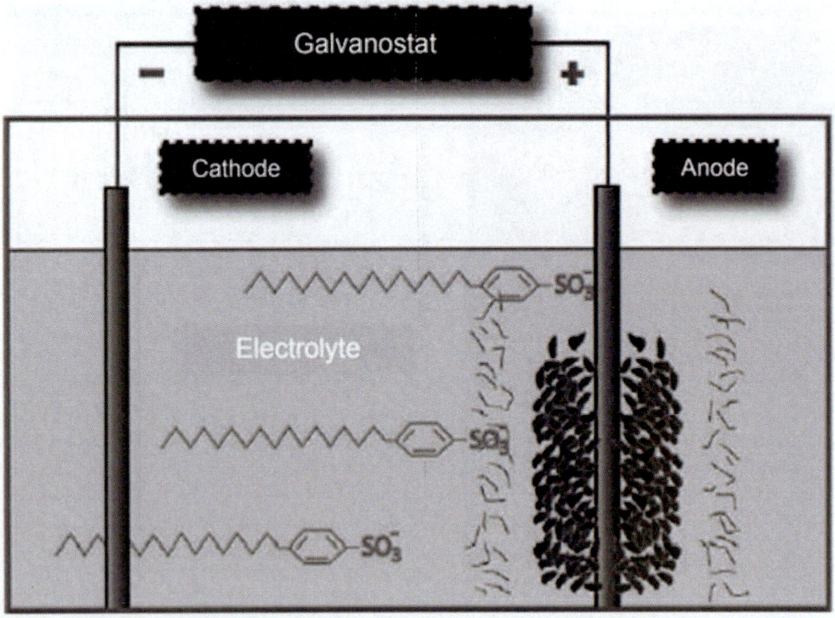

Fig. 10 Electrochemical exfoliation is used to synthesise graphene [43]

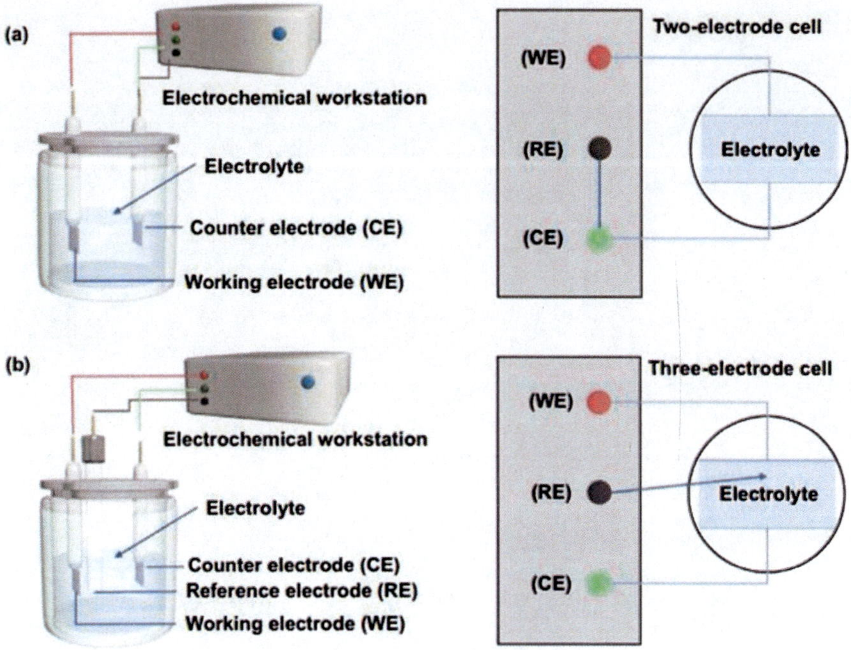

Fig. 11 Images depicting the experimental settings for electrochemical syntheses using a two-electrode and a three-electrode electrolytic cell [46]

at the edge.

$$H_2O - e^- \rightarrow OH^{\cdot} + H^+ - e^- \rightarrow O^{\cdot} + H^+$$

Then, the graphite anode expanded visibly in Stage II. Oxidation allowed the anionic BF_4^- to intercalate, resulting in the depolarization and expansion of the graphite anode.

$$C_x + BF_4^- \rightarrow BF_4C_x + e^- + H_2O \rightarrow C_xOH + HBF_4$$

Finally, expanded flakes peeled away from the anodes, forming a black slurry in the electrolyte solution in Stage III. Graphene nanoribbons were formed by oxidative cleavage of expanded graphene sheets, and some expanded sheets precipitated as graphene sheets. This approach has opened the door to large-scale manufacturing of fluorescent and biocompatible carbon nanomaterials for use in biological labelling and imaging.

Using the lithium rechargeable battery idea, Wang et al. demonstrated a solution approach for generating few-layer graphene flakes from negative-graphite electrodes [75]. Specifically, the co-intercalation of the electrolyte with Li-ions during electrochemical charging in a graphite electrode would generate a ternary graphite

Fig. 12 During exfoliation in 60 wt.% water/[BMI_m][BF_4] electrolyte, time evolution of IL electrolyte and highly oriented pyrolytic graphite (HOPG) anode. Panels (**b**), (**c**), and (**d**) depict Stages I, II, and III, respectively. The greatly expanded HOPG is illustrated in panel (**f**) [44]

intercalation compound at the graphite interlayers. Owing to the stress created by such ternary graphite intercalation compounds, the graphite would be eventually fragmented. Figure 13 depicts the schematic diagram. The graphene flakes were investigated to be roughly 1.5 nm thick.

The Raman spectra of dried graphene nanosheet powder is shown in Fig. 14, along with that of a graphite rod as an inset [73]. The Raman spectrum of graphene nanosheet powder revealed a broad D band (1350 cm^{-1}) and a strong G band (1580 cm^{-1}). The G line represents the in-plane bond-stretching motion of pairs of C sp^2 atoms or E_{2g} phonons, while the D line represents the breathing modes of rings or k-point phonons with A_{1g} symmetry [17]. The defects and partially disordered crystal structure in graphene nanosheets have been observed to produce a strong D band in the Raman spectrum, characterised by a high intensity ratio of $I_D/I_G > 1$ [42]. As the intensity of the G band is substantially higher than that of the D band in Fig. 14, the electrochemically exfoliated graphene nanosheets have a low fault content.

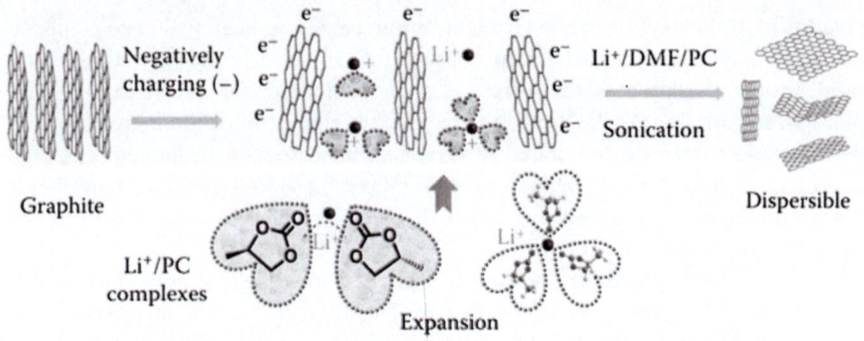

Fig. 13 Mechanism for exfoliation of graphite into few-layer graphene flakes via electrochemical intercalation of Li^+ complexes [75]

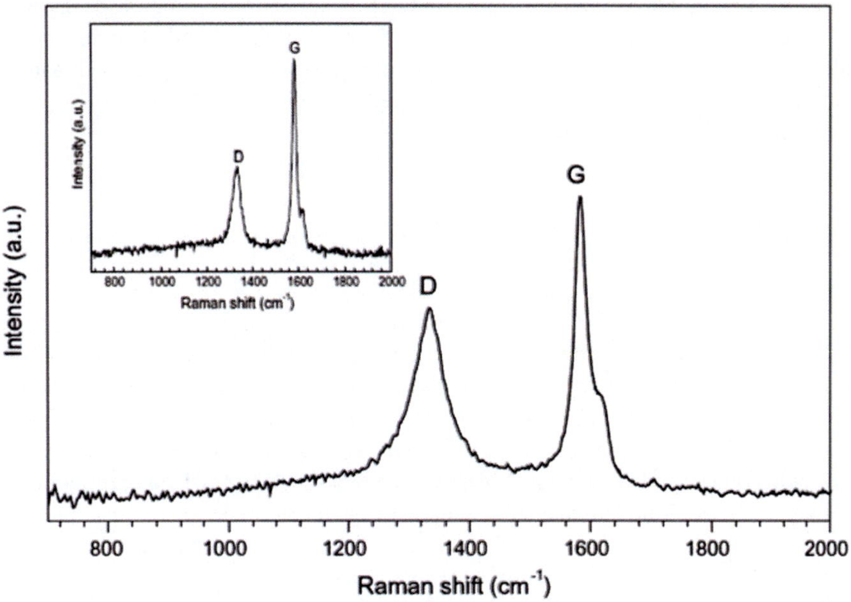

Fig. 14 Raman spectrum of the graphene nanosheet powders, with pristine graphite rod's Raman spectrum displayed in the inset [73]

Electrophoresis deposition can be employed to create graphene films from the graphene/SDS suspensions, as demonstrated in Fig. 15 [1]. Electrophoresis was carried out at 3 V for 3 h, after which the cathode and anode Pt electrodes were rinsed with ultra-pure water and dried with a stream of N_2. Figure 15a and b show low and high magnification SEM images of the Pt surface, respectively. Similarly, Fig. 15c and d show low and high magnification SEM images of the Pt anode electrode after the electrophoresis process. Before and after the electrolysis, the surface structure of the Pt anode remained essentially unchanged, indicating that no graphene film was produced on the surface. In contrast, Fig. 15e clearly shows that following the electrophoresis process, the surface of the Pt cathode was coated with nano- and micro-sized particles. As illustrated in Fig. 15f, a high magnification image revealed that the particles were agglomerated into graphene nanosheets with diameters as large as 1.3 μm. Electrochemical reduction of GO flakes produced graphene nanosheets that are similar to GO films electrochemically deposited on glassy carbon electrodes [6].

TEM images of graphene nanosheets treated at intercalation potentials of 1.4, 1.6, 1.8, and 2.0 V are shown in Fig. 16 [1]. The intercalation potential has a clear impact on the graphene nanosheets' transmission and form. The graphene nanosheets revealed multi-layered architectures in most cases. The sample processed at the intercalation potential of 1.4 V resulted in disordered and relatively thick graphene flakes, whereas the intercalation potential of 1.6 V resulted in comparatively transparent, more ordered, much larger and thin graphene flakes. Highly ordered hexagonal graphite lattices of multi-layered graphene flakes were discovered at higher intercalation potentials of 1.8 and 2.0 V, as shown in Fig. 16c and d. The thermodynamic stability of graphene's 2-D structure was linked to the buckled or wrinkled regions visible in Fig. 16c and d [50].

5 Arc Discharge

Wu et al. [79] reported the mechanism for the synthesis of high-quality graphene sheets in their research, as depicted in Fig. 17. To begin the process, the alternating current is used to cause two electrodes to consume and evaporate at the same time, preventing the formation of cathode deposits that occur in direct current arc-discharge processes [32]. Due to the abilities for the two electrodes to alternately act as anode and cathode, the arc-discharging zone between them is able to reach an exceptionally high temperature. This indicates that the rate of diffusion of carbon atoms, clusters, and gas molecules such as N_2 or H_2 around the arc-discharge zone would increase, reducing the likelihood of collision between all carbon species and gas molecules like N_2 or H_2. The distance between the arc-discharge area and the only source of cooling (the furnace wall) is significantly greater than the distance between two electrodes, resulting in a mild temperature gradient. Furthermore, the high thermal conductivity of the buffer gas, such as H_2, results in a sharp temperature gradient of plasma in the furnace. Since hydrogen gas cools quickly, carbon clusters form quickly before atoms deposit to create a crystal structure [84]. These clusters lack the energy and time to form a long-distance structured crystalline, so they form a disordered structure similar to amorphous carbon or thick graphene sheets. Inert gas having low thermal conductivity, such as N_2, will be used to make up for the shortfall. N_2 is regarded to be an important component in the synthesis of single-walled carbon nanohorns (SWCNHs) because it allows for the formation of curved structures by doping N atoms into the lattice and changing the temperature gradient [64]. On the

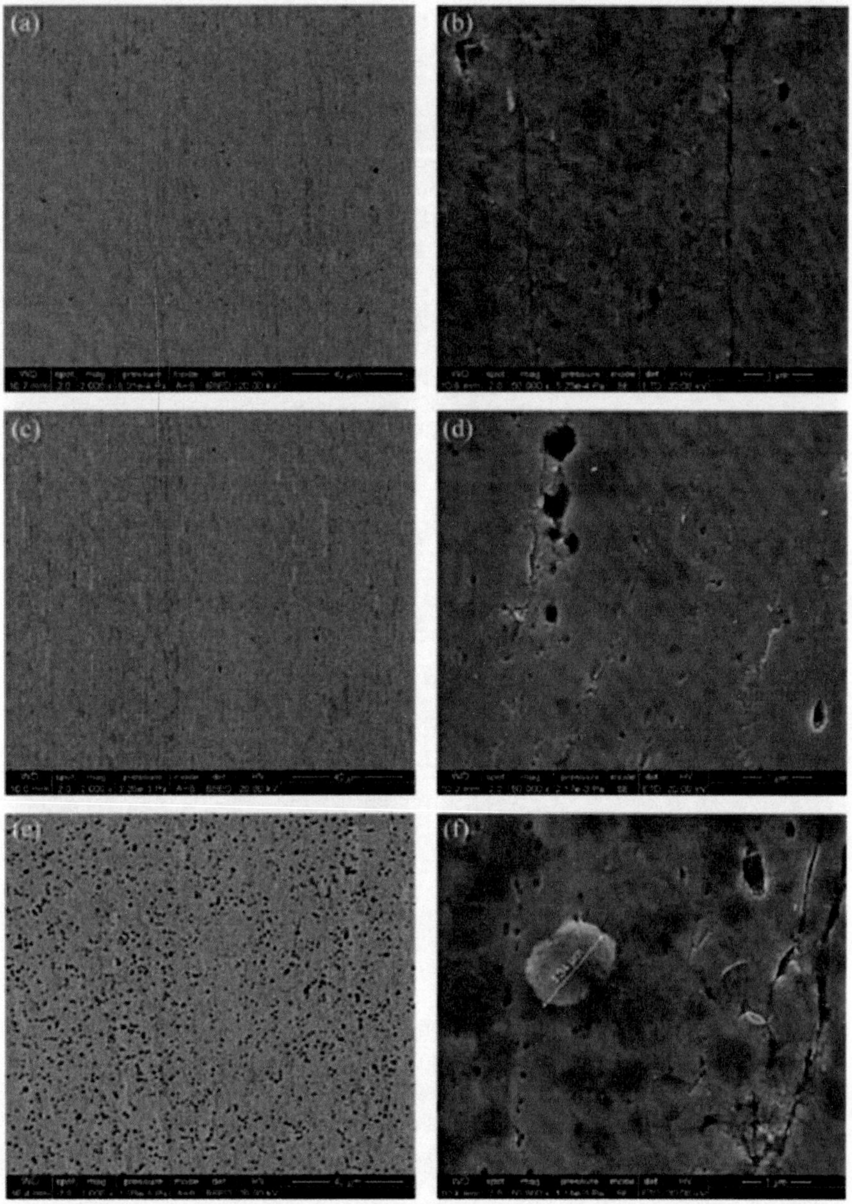

Fig. 15 SEM pictures of the Pt surface before (**a, b**) and after (**c, d** at anode and **e, f** at cathode) the electrophoresis process in the graphene/SDS suspension at 3 V for 3 h [1]

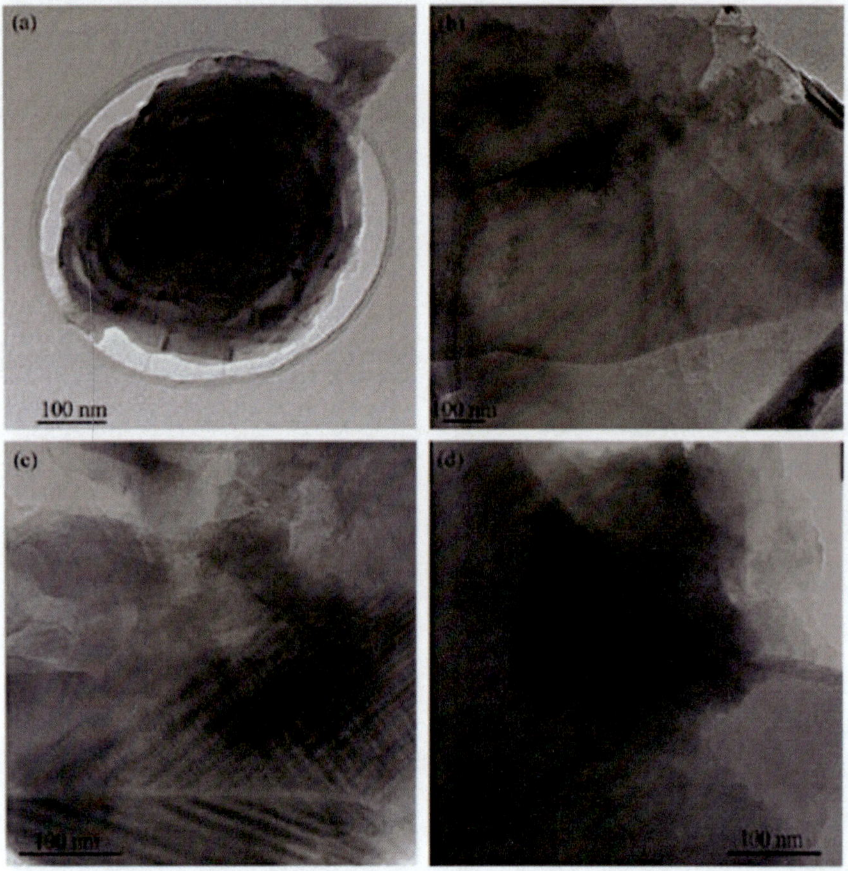

Fig. 16 **a** 1.4, **b** 1.6, **c** 1.8, and **d** 2.0 V. TEM images of graphene nanosheets generated at various intercalation potentials [1]

other hand, the presence of H_2 can effectively prevent the formation of dangling bonds and closed structures. As a result, graphene sheets made with the right ratio of N_2 to H_2 have a stable structure with few heteroatoms.

In H_2/He, direct current (DC) arc discharge between graphite electrodes produced graphene flakes that range from 2 to 4 layers, as well as multiwalled carbon nanotubes (MWCNTs) in the chamber wall [65]. Li et al [40] produced N-doped multilayer graphene in a He and NH_3 environment. By adjusting gas pressures and currents during arc discharge in a helium atmosphere, graphene sheets with various number of layers have been obtained. Using a maximum open circuit voltage of 60 V, the discharge current obtained varied between 100 and 150A. The arc was sustained by moving the cathode so that it was always 2 mm away from the anode. The production of N-doped graphene flakes was observed after an arc discharge was performed in a mixed mixture of He and NH_3 at 0.1 MPa pressure [40]. High-resolution TEM

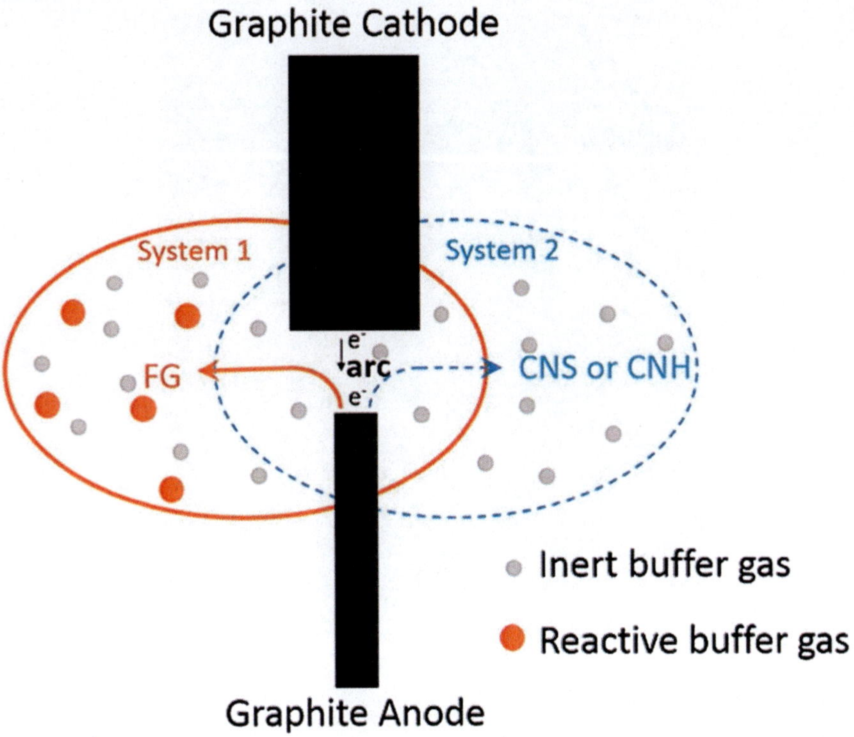

Fig. 17 Schematic diagram of the production of graphene sheets in various conditions. The N element is represented by blue atoms, while the C element is represented by black atoms [59]

examinations suggested that the flakes obtained in Fig. 18 were made up of 2–6 layers, and the N content was around 1%. Multilayer graphene flakes were synthesized by the means of external magnetic field-assisted electric arc discharge, carbon vapour sublimated from a graphite anode in an Ar atmosphere deposited on a graphite cathode and deposits consisting primarily of graphene flakes and carbon nanotubes (CNTs). The discharge was carried out with a 22 V arc voltage, 170 A arc current, and a 0.07 MPa Ar gas pressure. Graphene flakes containing few CNTs and carbon nanoparticles were created by setting the distance between the electrodes to 60 mm.

Graphene has also been synthesized by arc evaporation of graphite in the presence of hydrogen by Subrahmanyam et al. [65]. Graphene sheets with two to three layers and flake sizes of 100–200 nm are produced. This is achieved by utilizing the presence of H_2 during the arc discharge process to terminate dangling carbon-hydrogen bonds and preventing the creation of closed structures. High current (>100 A), high voltage (>50 V), and high hydrogen pressure (above 200 Torr) are all favourable conditions for producing graphene in the inner walls. Figures 19a and Fig. 19b show TEM and AFM pictures of the HG sample, respectively. This approach has been used to dope graphene with boron and nitrogen successfully [57]. The discharge is carried out in

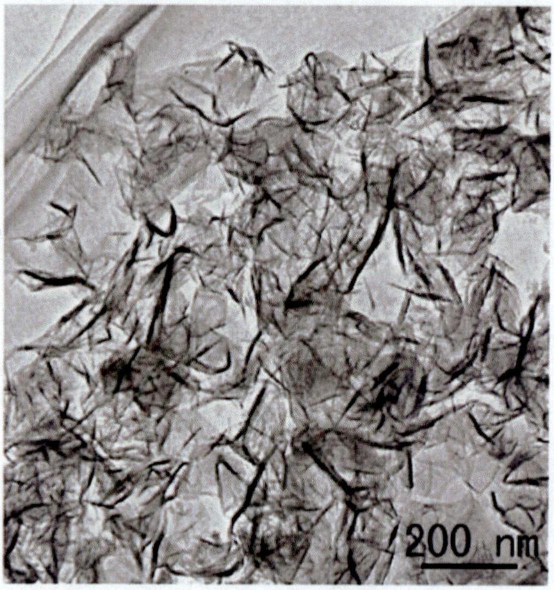

Fig. 18 Direct current arc discharge in He/NH_3 produced N-doped graphene flakes [40]

the presence of H_2 + diborane and H_2 + pyridine or ammonia, respectively, to prepare boron-doped graphene and nitrogen-doped graphene. Following these findings, some changes to the synthetic conditions resulted in the production of multilayer graphene on a large scale. Arc discharge in an air atmosphere produced graphene nanosheets with two layers that are 100–200 nm wide. The yield is highly influenced by the initial air pressure [76].

The Raman spectra of grown material in various electrodes are displayed in Fig. 20a [66]. The type of the synthesised materials was determined using the D-peak, 2D-peak and G-peak. Around 1350 cm^{-1} in Raman shift is the D peak for graphene, which represents the activation by defects. Other bands are present in graphene with defects, and these bands are triggered because of the crystal's symmetry being broken, which relaxes the Raman fundamental selection rule [49]. Such a phenomenon leads to an appearance of a D peak. The intensity of the 2D peak (~2700 cm^{-1}), which is derived from inelastic scattering of two phonons, and the G peak (~1580 cm^{-1}), which is related to the ordered in-plane sp^2 carbon structure, can be used to distinguish the number of graphene layers (mono-, bi-, or few-) [18]. Only a resonant one may change the Raman cross section for the double resonance mechanism at the peak about 2700 cm^{-1}. A new photon is eventually released when the excited electron is scattered back because of a phonon or a flaw [49]. Figure 20b and c indicate that the produced material in the cathode (the pink line in Fig. 20a) can be identified as graphene in a few layers. The amount of graphene layers affects the 2D peak's form and intensity. Figure 20b's inset partly enlarged depiction of the 2D peak looks to be symmetrical, and the symmetry axis value is around 2690 cm^{-1} (a little

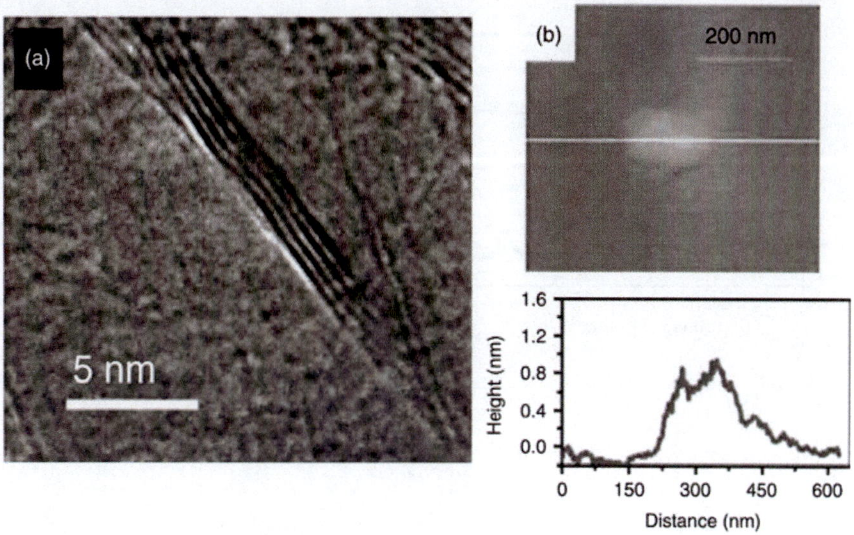

Fig. 19 HG produced by arc discharge of graphite in hydrogen **a** TEM and **b** AFM images. A height profile for the same can be found below [65]

less than 2700 cm^{-1}). The 2D peak downshifted as the number of graphene layers increased (from a monolayer to multilayers) [54]. When it comes to graphite, the 2D peak's shape—which can be thought of as an assembly of the $2D_1$ and $2D_2$ peaks—is not symmetrical and it exceeds 2700 cm^{-1} [18]. The cathode's grown products' 2D and G peaks have an intensity ratio of around 0.52. As illustrated in Fig. 20c, Tu and his colleagues verified that the layer of graphene changes from three to seven layers when the ratio is approximately 0.5 [70]. Additionally, Fig. 20a shows that the anode's (green line) produced material is graphite in thin layers since the I_{2D}/I_G intensity ratio is less than 0.5. The defect level is evaluated using the I_D/I_G intensity ratio. The anode's I_D/I_G ratio is higher than 0.7 in comparison to the cathode, indicating more defects.

6 Epitaxial Growth

In specifically built air-tight furnaces, graphene on SiC is grown at ambient pressure. Quartz tubes or vacuum chambers are used in the system. Since heat can be lost to the surrounding more effectively via convection at ambient pressure, appropriate water cooling of the furnace wall is required. To tolerate the high annealing temperature, a graphite holder is utilised. Figure 21 depicts a water-cooled quartz tube furnace that uses inductive heating. When using wafer size SiC samples, methods to reduce the temperature gradient must be addressed. At a rate of 2–3°Cs^{-1}, the SiC samples are slowly heated and cooled. The annealing temperature is maintained at 1500–2000 °C

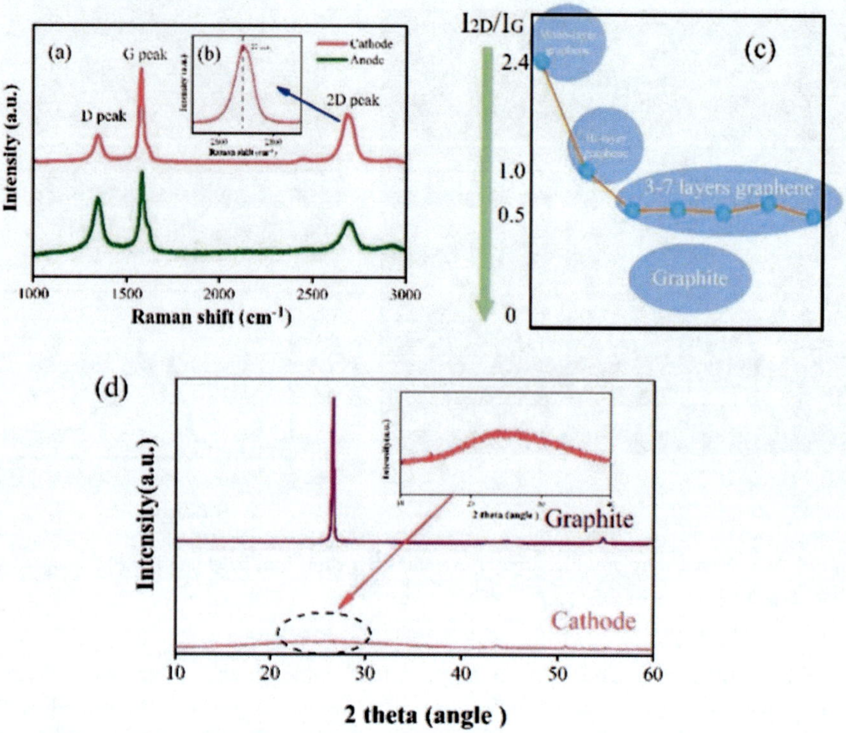

Fig. 20 Raman spectra of the synthesised material **a** in two electrodes, **b** a partial enlarged drawing of a 2D peak (inset), **c** a schematic diagram of the I_{2D}/I_G intensity ratio, **d** XRD patterns [66]

for 15 min. During the growth on Si-face SiC, argon flow at 0.9–1 bar is added into the furnace. To improve the suppression of quick silicon sublimation at the surface, the annealing temperature is fixed at 1500 °C and an argon pressure of up to 9 bar is utilised [72].

This method can produce a single-layer or multilayer graphene, as shown in Fig. 22. Silicon carbide (SiC) can be utilised as a precursor in this procedure, and graphene can be formed on a substrate using a high-temperature sublimation growth method. High temperatures (1300–1800 °C) and ultrahigh vacuum (UHV) or inert atmosphere are the synthesis conditions in this technique. Although, carbon has a higher sublimation temperature than silicon, silicon sublimes from the SiC surface at 1500 °C. As a result, a thin carbon layer can be formed [29].

In a graphene layer, the C atoms' surface density (3.82×10^{15} cm^{-2}) is three times that of a SiC bilayer (1.22×10^{15} cm^{-2}). This indicates that three SiC bilayers are consumed in order to generate a graphene layer. An established method is depicted in Fig. 23. A 0.07 nm high step and a 0.25 nm high step can be seen in the line profile in Fig. 23b taken along line AD in the STM image in Fig. 23a. A SiC bilayer process is responsible for the 0.25 nm high step. The model shown in Fig. 23c depicts the atomic

Fig. 21 A quartz tube furnace that uses inductive heating and water cooling to grow graphene in an Ar environment [12]

structures of nearby single-layer and bilayer epitaxial graphene (EG). A single SiC bilayer thermally decomposes beneath the interfacial graphene (IG) of a single layer EG, with Si species sublimating from the interface and carbon species releasing to form a new IG layer. This causes the original IG layer to be transformed into a new first EG layer atop the newly generated IG, culminating in the shift from single-layer to bilayer EG. The top EG layers of the neighbouring bilayer and single-layer EG remain continuous in this bottom-up growth paradigm because they originate from the same EG layer. This explains the physical continuity found at the single-layer-bilayer EG interface. The development of a second EG layer with interlayer spacing of 0.34 nm compensates for the lowering of the EG layer due to the disintegration of the underlying SiC bilayer, which is compatible with the measured height difference between the monolayer and bilayer EG which is 0.07 ± 0.01 nm [25].

Raman spectroscopy can be used to characterise the EG produced on SiC. For single-layer and bilayer EG on 6H-SiC(0001), single-layer mechanically cleaved graphene (MCG), bulk graphite, and bare 6H-SiC(0001) substrate, the typical Raman spectra are illustrated in Fig. 24 [53]. The bulk SiC related peaks arise at ~1520 and ~1713 cm^{-1} on the single-layer or bilayer EG, just as they do on the bare SiC substrate. The defect-induced D band at ~1368 cm^{-1}, the in-plane vibrational G band at ~1597 cm^{-1}, and the two-phonon 2D band at ~2725 cm^{-1} are three further peaks associated to EG. The insertion clearly demonstrates that the 2D band of

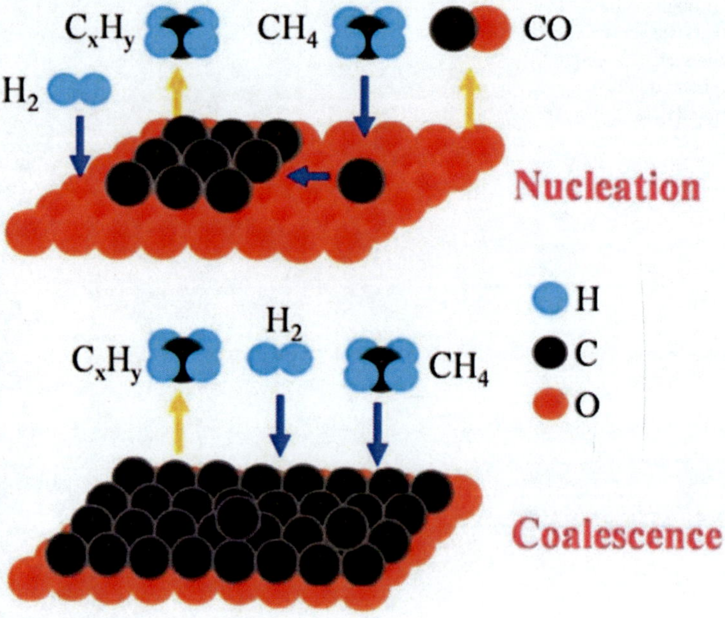

Fig. 22 Si sublimation on SiC surface [29]

bilayer EG is broader than that of single-layer EG (95 vs. 60 cm^{-1}) and occurs at a higher frequency (2736 vs. 2715 cm^{-1}), which is consistent with the MCG trend. The substantial lattice mismatch between SiC (a = 3.07 Å) and graphene (a = 2.46 Å) causes a significant blue shift of the G (1597 cm^{-1}) and 2D (2715 cm^{-1}) bands of single-layer EG compared to single-layer MCG (G band at 1580 cm^{-1} and 2D band at 2673 cm^{-1}).

Raman spectroscopy was also used to investigate epitaxial graphene produced on SiC substrate [53]. Si-terminated 6H-SiC(0001) was annealed at 1300 °C numerous times after being heated at 850 °C for 2 min in ultrahigh vacuum under a silicon flux. Graphene was synthesised on C-terminated 6H-SiC(0001) by annealing under the same circumstances without the use of a silicon flux. As illustrated in Fig. 25a, significant blue-shifts of the Raman bands D, G, and G' were found from the bands for both bulk graphite and micromechanically split graphene, which were contributed by the compressive strain generated by the SiC substrate. As demonstrated in Fig. 25b, double-layered graphene on C-terminated SiC exhibits a stronger D-band and a somewhat larger G-band, indicating a lower crystallinity than those on Si-terminated SiC. As the thickness of the epitaxial graphene layer increases, all Raman bands shift to lower frequencies, as illustrated in Fig. 25c. The influence of the substrate on epitaxial graphene weakens as the thickness of the graphene increases, implying that the graphene lattice relaxes.

To determine the thickness distribution, LEEM was used to characterise the graphene samples produced under similar circumstances on substrates made of 4H,

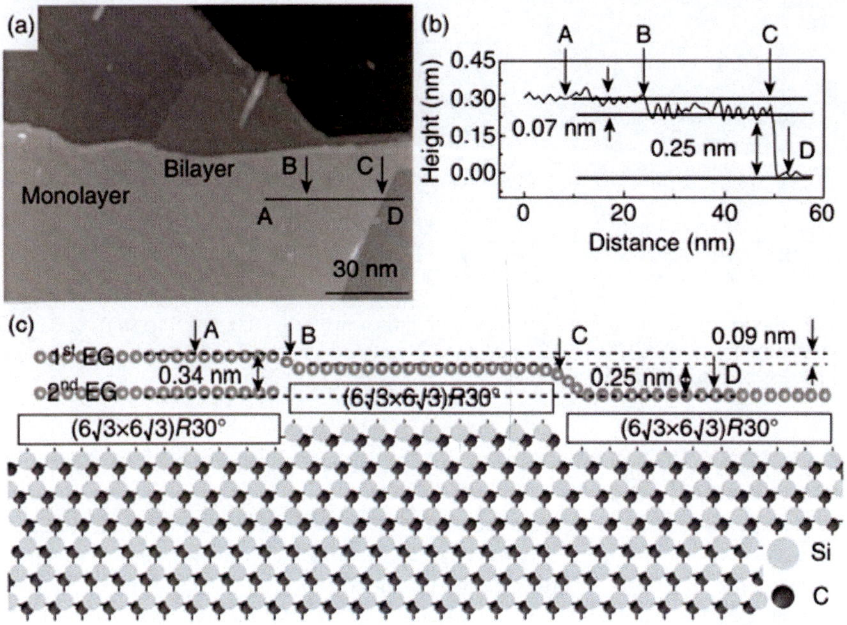

Fig. 23 **a** large-scale STM image of epitaxial single-layer and bilayer graphene on 6H-SiC(0001) ($150 \times 100\ nm^2$, $V_T = 1.5$ V); **b** line profile along the line AD in the image. **c** Schematic diagram of production of graphene via epitaxial growth [25]

Fig. 24 Raman spectra of single-layer and bilayer EG on SiC in comparison to bulk graphite and single-layer MCG [53]

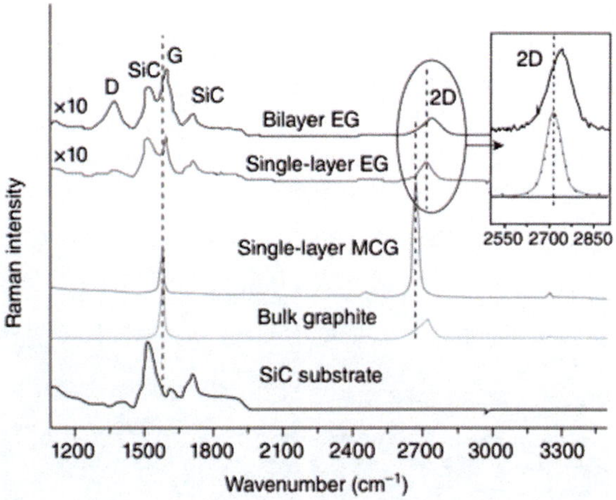

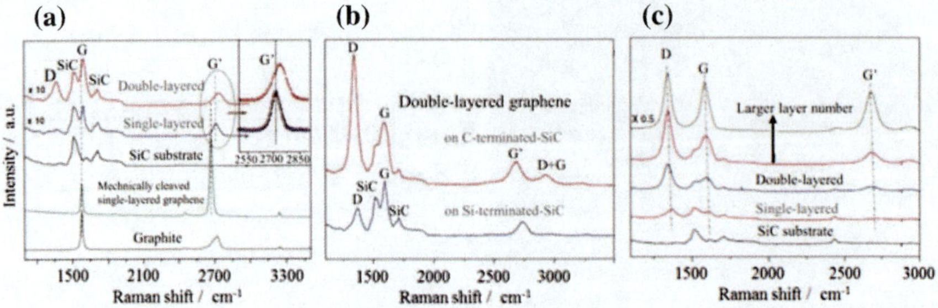

Fig. 25 Raman spectra of epitaxial graphene films grown on 6HeSiC(0001) **a** Blue shifts of Raman bands due to substrate limitation **b** Effect of the SiC crystal face, either C- or Si-terminated **c** Change in thickness of graphene sheets generated [53]

6H, and 3C-SiC [83]. A multi-layered (ML) graphene is represented by the bright region in each LEEM picture (Fig. 26), whereas bilayer graphene is represented by the darker portions. Graphene grown on a 4H-SiC substrate's LEEM image is depicted in Fig. 26a. One and two MLs of graphene, with a little region of three MLs, are present in this sample (the black spot). Figure 26b and c display large, homogenous ML graphene grown on 6H-SiC and 3C-SiC. For the 4H, 6H and 3C polytypes, respectively, the extracted regions with 1ML coverage are roughly 60, 90, and 98% from the LEEM images.

Graphene grown on unpolished (as-grown) and polished 3C-SiC (1 1 1) substrates are seen in LEEM images in Fig. 27a and Fig. 27b, respectively [83]. On an unpolished substrate, graphene makes up 65% of the bilayer, with three MLs covering certain places (dark area). One ML of graphene covers 93% of the overall surface (bright area) of the polished substrate, with some tiny places having a graphene thickness of two MLs (Fig. 27b). Figure 27c and Fig. 27d, respectively, show AFM topography images of graphene on the as-grown 3C-SiC substrate and on the polished sample

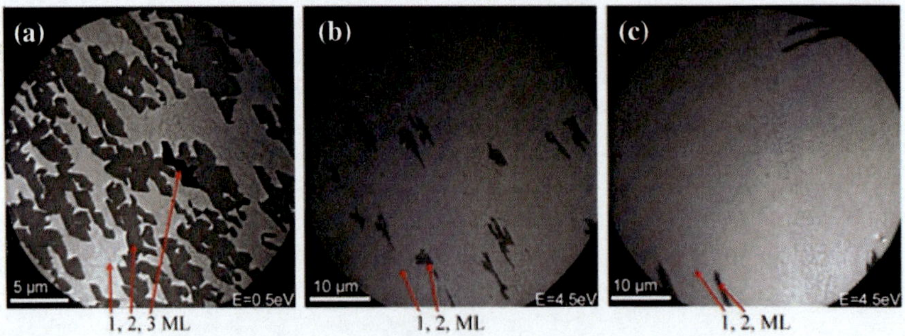

Fig. 26 LEEM images of graphene on **a** 4H-SiC with 60% ML coverage (bright area), where darker regions represent bilayers and the little black spot represents three layers of graphene **b** 6H-SiC with 92% ML coverage (bright area), **c** 3C-SiC with 98% ML coverage (bright area). In (**b**) and (**c**), two ML graphene is represented by the black spots [83]

together with their corresponding roughness parameters (rms). The latter sample's angle-resolved photoemission spectroscopy (ARPES) spectra of the pi band, which was recorded at the K point of the graphene Brillouin zone, exhibits good linear dependency typical of 1 ML graphene. The polished and unpolished samples had a surface roughness of 2 nm and 0.6 nm, respectively. A decreased surface roughness leads to less noticeable step bunching with less variation of step heights, which improves the quality of the generated graphene, as seen in Fig. 27e and f. Therefore, surface roughness needs to be kept to a minimum before graphene formation.

7 Pyrolysis

Graphene is prepared by pyrolyzing a poly(methyl methacrylate) (PMMA) composite in 6 steps, as shown in Fig. 28(1)–(6). PMMA decomposes into numerous carbon derivatives (C_n) when the composite is heated to 1000 °C. . Carbon derivatives are inhibited from evolving in the limited microzones produced by organophilic montmorillonite (OMT) due to the presence of OMT layers. After the addition of OMT layers, the carbon derivatives will have a longer time to interact with catalysts, resulting in a greater char [67]. Carbon atoms (C) from carbon derivatives dissolve quickly on the Ni fcatalyst surface, which has a high carbon solubility at a high temperature. Carbon atoms start to precipitate from the hexagonally close packed Ni surfaces until the solution becomes oversaturated, resulting in precisely ordered hexagonal rings [77]. Since micro grade Ni particles have a huge surface area, they could be employed to substitute the flat metal substrate as active sites for graphene synthesis. The growth process of graphene on Ni microparticles is close to that of graphene on a flat Ni surface, hence graphene is formed through the epitaxial growth of hexagonal rings on the Ni surface [10]. On the other hand, the clay layers move to the surface, forming a charred ceramic coating that protects the surface. After removing the clay and Ni particles, pure graphene can be obtained. However, compared to Cu catalyst, Ni catalyst has a certain drawback, which being the carbide layer would be formed at high temperature such as 1000 °C [56].

The XRD pattern of the produced graphene was shown in Fig. 28a to further describe the structures. The usual (0 0 2) plane of hexagonal graphite is shown by a prominent diffraction peak at 26.6°. The spacing between two neighbouring graphene sheets is around 0.34 nm, according to the diffraction peak location of C(0 0 2), which agrees well with the HRTEM finding [80]. Furthermore, the presence of defect structures in graphene, which is derived from the structure of PMMA precursor, is indicated by a broad peak at around 21.4°. In addition, the 2θ peak, which is connected to the 2D in-plane symmetry along the graphene sheets [14], can be seen a roughly 43.1° (1 0 0). Since it is sensitive to tiny changes inside the carbon layers, Raman spectroscopy was utilised to study the crystallinity of carbon materials [58]. The Raman spectroscopy of the produced graphene is shown in Fig. 28b. The D and G bands are represented by strong peaks in the Raman spectra at 1354 cm^{-1} and 1581 cm^{-1}, respectively. Amorphous carbon and lattice imperfections in graphene

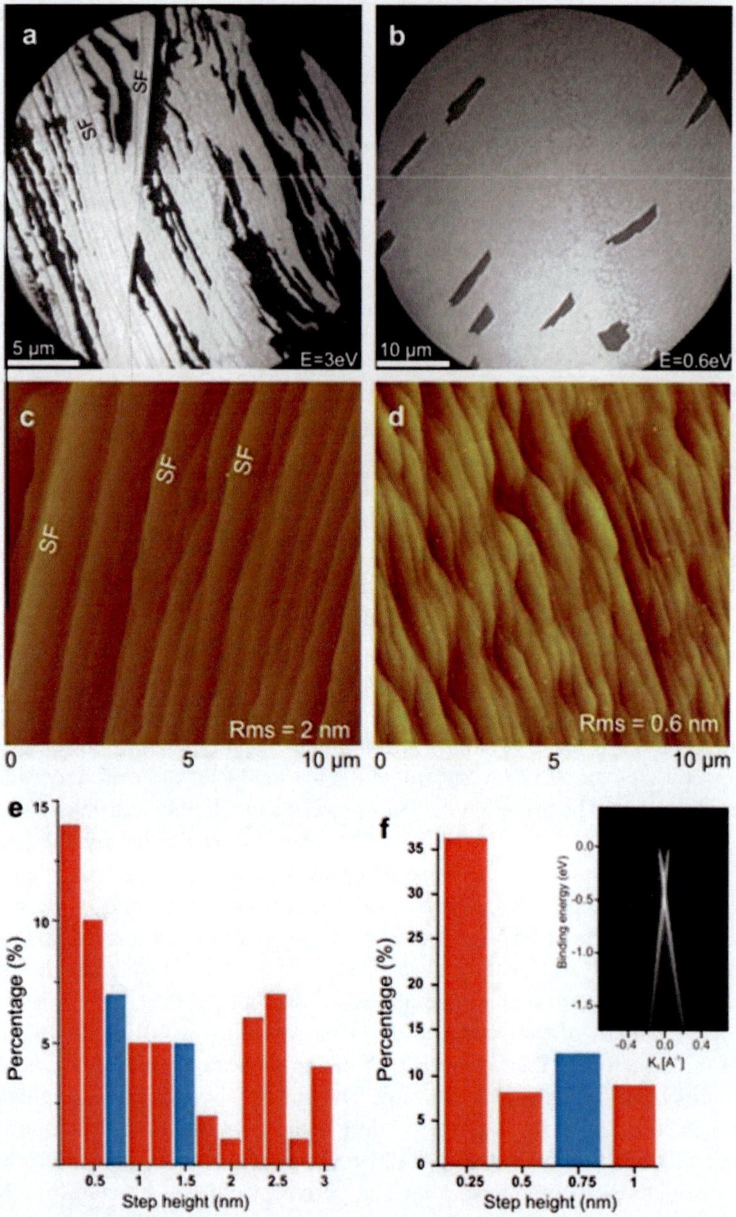

Fig. 27 LEEM images of the graphene layers on **a** as grown 3C-SiC with ~65% coverage by 2 ML graphene (bright region) and the remaining ~35% by 3 ML, along with some stacking faults (SF) as well as **b** polished 3C-SiC with ~7% of 2 MLs and ~93% covered by an ML graphene (bright region). AFM images of graphene on as grown 3C-SiC and substrates made of polished 3C-SiC are presented in (**c**) and (**d**), respectively. A wide range of step heights are shown in the step height histograms for **e** graphene on a grown substrate and **f** graphene on a polished substrate. ARPES spectrum of the p band captured at the K point is shown in the inset image [83]

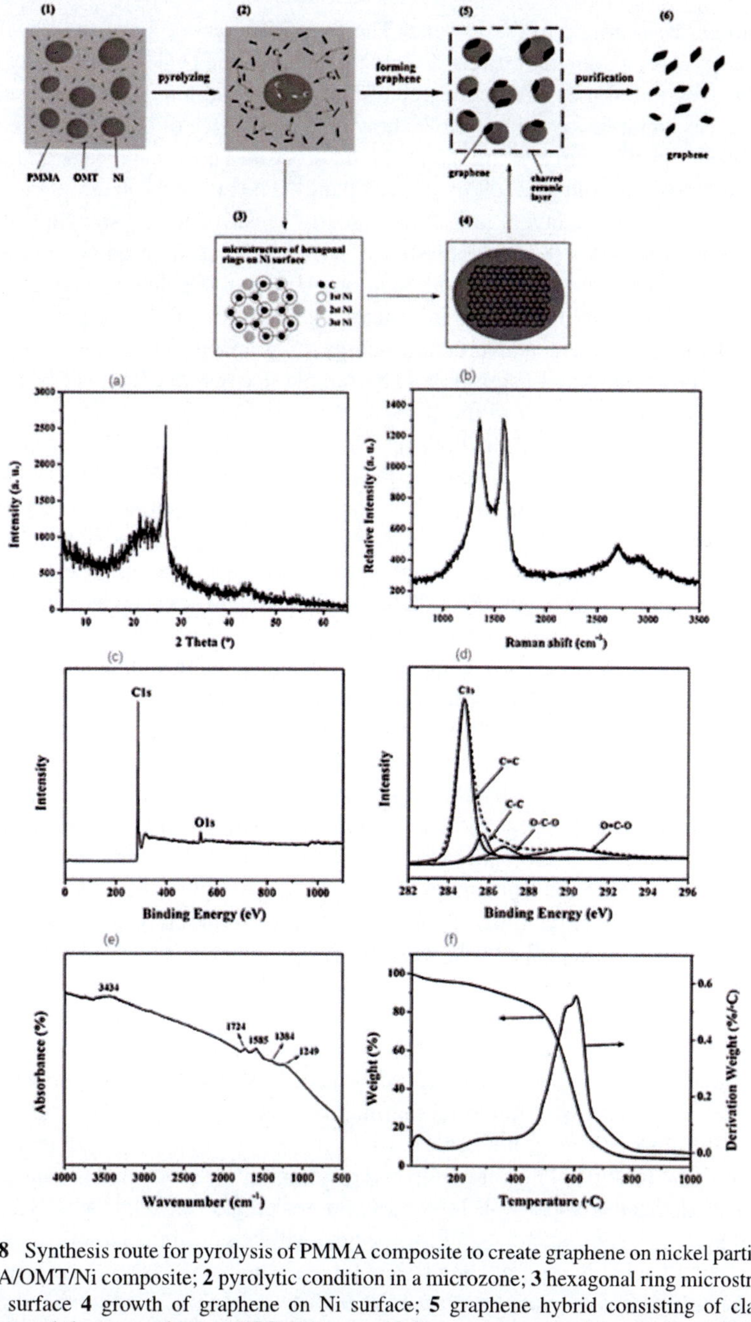

Fig. 28 Synthesis route for pyrolysis of PMMA composite to create graphene on nickel particles: **1** PMMA/OMT/Ni composite; **2** pyrolytic condition in a microzone; **3** hexagonal ring microstructure on Ni surface **4** growth of graphene on Ni surface; **5** graphene hybrid consisting of clay and catalyst; and **6** pure graphene; **a** XRD pattern and **b** Raman spectrum of the graphene; Full-scale XPS spectrum **c** higher resolution curves of C1s **d** FTIR spectrum **e** TGA curve **f** of the obtained graphene [23]

sheets might be ascribed to the D band. The first-order scattering of the $E^2{}_g$ mode from the sp^2 carbon domains corresponds to the G band. The D/G band intensity ratio is frequently used to measure the degree of disorder in graphite [36]. The resulting graphene has an intensity ratio of 1.67, showing the presence of chemical defects in the graphene manufactured using the pyrolytic process. The D band's second order is the wide peak at around 2709 cm^{-1} (2D band). The strength of the 2D band is related to the graphene layers and is particularly sensitive to the stacking of one another along the c-axis [82]. The existence of extremely disordered and randomly distributed graphene sheets with more than one layer is indicated by the broad 2D band, which is compatible with the information gained from TEM images.

The elemental content and bonding energy of graphene generated using this approach were determined using XPS. The sample is constituted of 94.6% carbon and 5.4% oxygen, according to the XPS spectrum in Fig. 28c, indicating the high carbon content. Figure 28d shows high-resolution C1s XPS spectra. The strong peak at 284.8 eV comes from the sp^2 carbon network in graphene (C = C) [9], which is greater than the sp^3-carbon (C–C) peak at 285.7 eV. Two further peaks, at 286.8 eV and 288.7 eV, are ascribed to the C–O–C and O = C–O groups, respectively, coming from epoxide, ether, and carboxyl groups in graphene [37]. The sp^2 -hybridized C bonding has a larger peak intensity than the C containing oxygen bonding, indicating that the generated graphene has a high degree of graphitization. The results of elemental analysis are mostly comparable with those of XPS, indicating that the samples contain mostly carbon atoms with minor amounts of oxygen and hydrogen. The chemical and structural information of graphene was investigated using FTIR. The FTIR spectra of the purified graphene produced by high-temperature pyrolysis of PMMA are shown in Fig. 28e. The presence of graphene structure is demonstrated by the distinctive strong peak at 1585 cm^{-1} attributed to the C = C stretching mode of the sp^2 C network. The FTIR spectrum shows a tiny C = O stretching at 1724 cm^{-1} produced from the original precursor. The O–H deformation, the C–O stretching peak, and the C–O–C stretching vibrations are found in the graphene produced by the pyrolytic technique [87]. Weak peaks for oxygen-containing functional groups are observed at 3434, 1384, and 1225 cm^{-1}. With both graphite and amorphous carbon, TGA was employed to assess the graphitic character of the graphene, shown in Fig. 28f. Under an air atmosphere, the graphene was heated at a rate of 20 °C/min from 50 to 1000 °C. As shown in Fig. 28f, weight loss of graphene occurs in three phases, resulting in three differential thermogravimetric (DTG) peaks. The evaporation of the absorbed water molecule causes graphene to deteriorate before 100 °C [80]. The disintegration of the remaining oxygen-containing functional groups (200–400 °C) in graphene is responsible for the second weak region of weight loss. Between 400 and 800 °C, a significant weight loss occurs, which is attributed to the oxidation of the carbon backbone of graphene sheets [40]. As high as 1000 °C, the resulting graphene still retains 4.2 weight percent. The TGA curve indicates that the number of functional groups on graphene sheets derived from the material is lower than that on graphene sheets generated by chemical techniques [15].

Digital pictures of graphene dispersed in ethanol at a concentration of 10 mg/L are shown in Fig. 29a. After sonication, graphene can be distributed sufficiently enough to create a stable suspension in ethanol for months, making graphene-based nanocomposites straightforward to make. Figure 29b shows the UV–vis absorption spectra of graphene. Due to p → p* transitions of aromatic C–C bonds in the polyaromatic system of graphene layers, graphene exhibits a high absorption peak at 265 nm [23].

The detonation reaction of a powder mixture of cyanuric chloride and trinitrophenol at 320 °C in autoclave produced N-doped graphene flakes with a momentary pressure of 60 MPa and an equilibrium pressure of 30 MPa [16]. Figure 30 shows TEM images and Raman spectra of the resulting flakes, indicating the production of thin multi-layered graphene flakes (4–8 layers stacked).

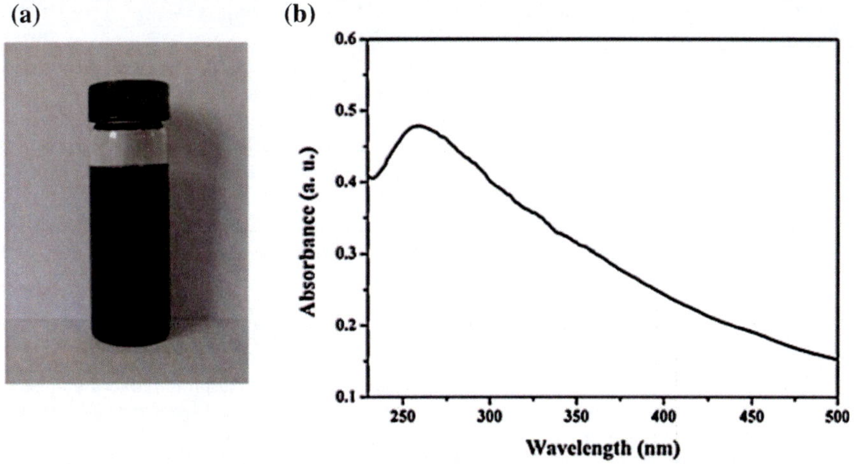

Fig. 29 **a** Typical photograph of dispersion of the as-prepared graphene at a concentration of 10 mg/L and **b** UV–vis absorption of the graphene [23]

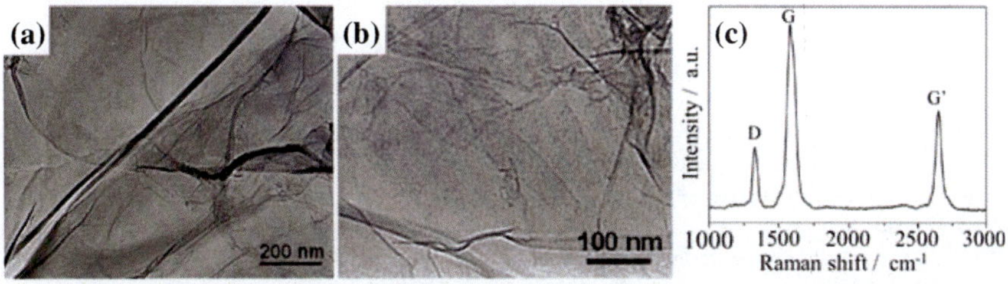

Fig. 30 **a** and **b** Transmission electron microscopy images; **c** Raman spectra of multilayered graphene flakes generated by detonation [16]

References

1. Alanyalioğlu, M., Segura, J.J., Oró-Sol, J., Casañ-Pastor, N.: The synthesis of graphene sheets with controlled thickness and order using surfactant-assisted electrochemical processes. Carbon **50**(1), 142–152 (2012). https://doi.org/10.1016/j.carbon.2011.07.064
2. Bae, S., Kim, H., Lee, Y., Xu, X., Park, J.S., Zheng, Y., Balakrishnan, J., Lei, T., Ri Kim, H., Song, Y.I., Kim, Y.J., Kim, K.S., Özyilmaz, B., Ahn, J.H., Hong, B.H., Iijima, S.: Roll-to-roll production of 30-inch graphene films for transparent electrodes. Nat. Nanotechnol. **5**(8), 574–578 (2010). https://doi.org/10.1038/nnano.2010.132
3. Berlouis, L.E.A., Schiffrin, D.J.: The electrochemical formation of graphitebisulphate intercalation compounds. J. Appl. Electrochem. **13**(2), 147–155 (1983). https://doi.org/10.1007/BF00612475
4. Blake, P., Hill, E.W., Castro Neto, A.H., Novoselov, K.S., Jiang, D., Yang, R., Booth, T.J., Geim, A.K.: Making graphene visible. Appl. Phys. Lett. **91**(6), (2007). https://doi.org/10.1063/1.2768624
5. Casallas Caicedo, F.M., Vera López, E., Agarwal, A., Drozd, V., Durygin, A., Franco Hernandez, A., Wang, C.: Synthesis of graphene oxide from graphite by ball milling. Diam. Relat. Mater. **109**(September), 108064 (2020). https://doi.org/10.1016/j.diamond.2020.108064
6. Chen, C., Zhai, W., Lu, D., Zhang, H., Zheng, W.: A facile method to prepare stable noncovalent functionalized graphene solution by using thionine. Mater. Res. Bull. **46**(4), 583–587 (2011). https://doi.org/10.1016/j.materresbull.2010.12.024
7. Chen, J., Cui, M., Wu, G., Wang, T., Mbengue, J.M., Li, Y., Li, M.: Fast growth of large single-crystalline graphene assisted by sequential double oxygen passivation. Carbon **116**, 133–138 (2017). https://doi.org/10.1016/j.carbon.2017.01.108
8. Chen, J.H., Jang, C., Xiao, S., Ishigami, M., Fuhrer, M.S.: Intrinsic and extrinsic performance limits of graphene devices on SiO_2. Nat. Nanotechnol. **3**(4), 206–209 (2008). https://doi.org/10.1038/nnano.2008.58
9. Chen, L., Tang, Y., Wang, K., Liu, C., Luo, S.: Direct electrodeposition of reduced graphene oxide on glassy carbon electrode and its electrochemical application. Electrochem. Commun. **13**(2), 133–137 (2011). https://doi.org/10.1016/j.elecom.2010.11.033
10. Chen, Z., Ren, W., Liu, B., Gao, L., Pei, S., Wu, Z.S., Zhao, J., Cheng, H.M.: Bulk growth of mono- to few-layer graphene on nickel particles by chemical vapor deposition from methane. Carbon **48**(12), 3543–3550 (2010). https://doi.org/10.1016/j.carbon.2010.05.052
11. Choy, K.L.: Chemical vapour deposition of coatings. Prog. Mater Sci. **48**(2), 57–170 (2003). https://doi.org/10.1016/S0079-6425(01)00009-3
12. De Heer, W.A., Berger, C., Ruan, M., Sprinkle, M., Li, X., Hu, Y., Zhang, B., Hankinson, J., Conrad, E.: Large area and structured epitaxial graphene produced by confinement controlled sublimation of silicon carbide. Proc. Natl. Acad. Sci. U.S.A. **108**(41), 16900–16905 (2011). https://doi.org/10.1073/pnas.1105113108
13. Di, C.A., Wei, D., Yu, G., Liu, Y., Guo, Y., Zhu, D.: Patterned graphene as source/drain electrodes for bottom-contact organic field-effect transistors. Adv. Mater. **20**(17), 3289–3293 (2008). https://doi.org/10.1002/adma.200800150
14. Dimovski, S., Nikitin, A., Ye, H., Gogotsi, Y.: Synthesis of graphite by chlorination of iron carbide at moderate temperatures. J. Mater. Chem. **14**(2), 238–243 (2004). https://doi.org/10.1039/b311938f
15. Economopoulos, S.P., Rotas, G., Miyata, Y., Shinohara, H., Tagmatarchis, N.: Exfoliation and chemical modification using microwave irradiation affording highly functionalized graphene. ACS Nano **4**(12), 7499–7507 (2010). https://doi.org/10.1021/nn101735e
16. Feng, L., Chen, Y., Chen, L.: Easy-to-operate and low-temperature synthesis of gram-scale nitrogen-doped graphene and its application as cathode catalyst in microbial fuel cells. ACS Nano **5**(12), 9611–9618 (2011). https://doi.org/10.1021/nn202906f
17. Ferrari, A.C., Robertson, J.: Interpretation of Raman spectra of disordered and amorphous carbon. Schweiz. Z. Für Hydrol. **31**(2), 632–645 (1969). https://doi.org/10.1007/BF02543692

18. Ferrari, A.C., Meyer, J.C., Scardaci, V., Casiraghi, C., Lazzeri, M., Mauri, F., Piscanec, S., Jiang, D., Novoselov, K.S., Roth, S., Geim, A.K.: Raman spectrum of graphene and graphene layers. Phys. Rev. Lett. **97**(18), 1–4 (2006). https://doi.org/10.1103/PhysRevLett.97.187401
19. Film, C., Ago, H., Ito, Y., Mizuta, N., Yoshida, K., Hu, B., Orofeo, C.M.: Epitaxial chemical vapor deposition growth of single-layer graphene over. Acsnano **4**(12), 7407–7414 (2010)
20. Guerrero-Fajardo, C.A., Giraldo, L., Moreno-Piraján, J.C.: Preparation and characterization of graphene oxide for Pb(II) and Zn(II) ions adsorption from aqueous solution: experimental, thermodynamic and kinetic study. Nanomaterials **10**(6), (2020). https://doi.org/10.3390/nano10061022
21. Hata, K., Futaba, D.N., Mizuno, K., Namai, T., Yumura, M., Iijima, S.: Water-assisted highly efficient synthesis of impurity-free single-walled carbon nanotubes. Science **306**(5700), 1362–1364 (2004). https://doi.org/10.1126/science.1104962
22. Hathcock, K.W., Brumfield, J.C., Goss, C.A., Irene, E.A., Murray, R.W.: Incipient electrochemical oxidation of highly oriented pyrolytic graphite: correlation between surface blistering and electrolyte anion intercalation. Anal. Chem. **67**(13), 2201–2206 (1995). https://doi.org/10.1021/ac00109a045
23. Hong, N., Yang, W., Bao, C., Jiang, S., Song, L., Hu, Y.: Facile synthesis of graphene by pyrolysis of poly(methyl methacrylate) on nickel particles in the confined microzones. Mater. Res. Bull. **47**(12), 4082–4088 (2012). https://doi.org/10.1016/j.materresbull.2012.08.049
24. Hu, B., Ago, H., Ito, Y., Kawahara, K., Tsuji, M., Magome, E., Sumitani, K., Mizuta, N., Ikeda, K.I., Mizuno, S.: Epitaxial growth of large-area single-layer graphene over Cu(1 1 1)/sapphire by atmospheric pressure CVD. Carbon **50**(1), 57–65 (2012). https://doi.org/10.1016/j.carbon.2011.08.002
25. Huang, H., Chen, W., Chen, S., Thye, A., Wee, S.: Bottom-up growth of epitaxial. ACS Nano **2**(12), 2513–2518 (2008)
26. Huang, L., Cui, X., Dukovic, G., O'Brien, S.P.: Self-organizing high-density single-walled carbon nanotube arrays from surfactant suspensions. Nanotechnology **15**(11), 1450–1454 (2004). https://doi.org/10.1088/0957-4484/15/11/012
27. Huang, P.Y., Ruiz-Vargas, C.S., Van Der Zande, A.M., Whitney, W.S., Levendorf, M.P., Kevek, J.W., Garg, S., Alden, J.S., Hustedt, C.J., Zhu, Y., Park, J., McEuen, P.L., Muller, D.A.: Grains and grain boundaries in single-layer graphene atomic patchwork quilts. Nature **469**(7330), 389–392 (2011). https://doi.org/10.1038/nature09718
28. Hummers W.S., Offeman, R.E.: Preparation of graphitic oxide. J. Am. Chem. Soc. **208**(1937), 1937 (1957). https://pubs.acs.org/sharingguidelines
29. Hwang, J., Kim, M., Campbell, D., Alsalman, H.A., Kwak, J.Y., Shivaraman, S., Woll, A.R., Singh, A.K., Hennig, R.G., Gorantla, S., Rümmeli, M.H., Spencer, M.G.: Van der waals epitaxial growth of graphene on sapphire by chemical vapor deposition without a metal catalyst. ACS Nano **7**(1), 385–395 (2013). https://doi.org/10.1021/nn305486x
30. Jayasena, B., Subbiah, S.: A novel mechanical cleavage method for synthesizing few-layer graphenes. Nanoscale Res. Lett. **6**(1), 1–7 (2011). https://doi.org/10.1186/1556-276X-6-95
31. Jeong, H.K., Yun, P.L., Lahaye, R.J.W.E., Park, M.H., Kay, H.A., Ick, J.K., Yang, C.W., Chong, Y.P., Ruoff, R.S., Young, H.L.: Evidence of graphitic AB stacking order of graphite oxides. J. Am. Chem. Soc. **130**(4), 1362–1366 (2008). https://doi.org/10.1021/ja076473o
32. Karmakar, S., Nawale, A.B., Lalla, N.P., Sathe, V.G., Kolekar, S.K., Mathe, V.L., Das, A.K., Bhoraskar, S.V.: Gas phase condensation of few-layer graphene with rotational stacking faults in an electric-arc. Carbon **55**, 209–220 (2013). https://doi.org/10.1016/j.carbon.2012.12.029
33. Kim, J., Ishihara, M., Koga, Y., Tsugawa, K., Hasegawa, M., Iijima, S.: Low-temperature synthesis of large-area graphene-based transparent conductive films using surface wave plasma chemical vapor deposition. Appl. Phys. Lett. **98**(9), 2011–2014 (2011). https://doi.org/10.1063/1.3561747
34. Kim, K., Artyukhov, V.I., Regan, W., Liu, Y., Crommie, M.F., Yakobson, B.I., Zettl, A.: Ripping graphene: preferred directions. Nano Lett. **12**(1), 293–297 (2012). https://doi.org/10.1021/nl203547z

35. Kim, K.S., Zhao, Y., Jang, H., Lee, S.Y., Kim, J.M., Kim, K.S., Ahn, J.H., Kim, P., Choi, J.Y., Hong, B.H.: Large-scale pattern growth of graphene films for stretchable transparent electrodes. Nature **457**(7230), 706–710 (2009). https://doi.org/10.1038/nature07719
36. Kim, N.S., Lee, Y.T., Park, J., Han, J.B., Choi, Y.S., Choi, S.Y., Choo, J., Lee, G.H.: Vertically aligned carbon nanotubes grown by pyrolysis of iron, cobalt, and nickel phthalocyanines. J. Phys. Chem. B **107**(35), 9249–9255 (2003). https://doi.org/10.1021/jp034895o
37. Lambert, T.N., Luhrs, C.C., Chavez, C.A., Wakeland, S., Brumbach, M.T., Alam, T.M.: Graphite oxide as a precursor for the synthesis of disordered graphenes using the aerosol-through-plasma method. Carbon **48**(14), 4081–4089 (2010). https://doi.org/10.1016/j.carbon.2010.07.015
38. Li, D., Kaner, R.B.: Materials science: graphene-based materials. Science **320**(5880), 1170–1171 (2008). https://doi.org/10.1126/science.1158180
39. Li, N., Zhen, Z., Zhang, R., Xu, Z., Zheng, Z., He, L.: Nucleation and growth dynamics of graphene grown by radio frequency plasma-enhanced chemical vapor deposition. Sci. Rep. **11**(1), 1–10 (2021). https://doi.org/10.1038/s41598-021-85537-3
40. Li, N., Wang, Z., Zhao, K., Shi, Z., Gu, Z., Xu, S.: Large scale synthesis of N-doped multi-layered graphene sheets by simple arc-discharge method. Carbon **48**(1), 255–259 (2010). https://doi.org/10.1016/j.carbon.2009.09.013
41. Li, X., Cai, W., An, J., Kim, S., Nah, J., Piner, R., Velamakanni, A., Jung, I., Tutuc, E., Banerjee, S.K.: Large-area synthesis of high-quality and uniform graphene films on copper foils. Science **324**(5932), 1312–1314 (2016)
42. Lotya, M., Hernandez, Y., King, P.J., Smith, R.J., Nicolosi, V., Karlsson, L.S., Blighe, F.M., De, S., Zhiming, W., McGovern, I.T., Duesberg, G.S., Coleman, J.N.: Liquid phase production of graphene by exfoliation of graphite in surfactant/water solutions. J. Am. Chem. Soc. **131**(10), 3611–3620 (2009). https://doi.org/10.1021/ja807449u
43. Low, C.T.J., Walsh, F.C., Chakrabarti, M.H., Hashim, M.A., Hussain, M.A.: Electrochemical approaches to the production of graphene flakes and their potential applications. Carbon **54**, 1–21 (2013). https://doi.org/10.1016/j.carbon.2012.11.030
44. Lu, J., Yang, J., Wang, J., Lim, A., Wang, S., Loh, K.P.: One-pot synthesis of fluorescent carbon graphene by the exfoliation of graphite in ionic liquids. ACS Nano **3**(8), 2367–2375 (2009)
45. Luo, B., Liu, H., Jiang, L., Jiang, L., Geng, D., Wu, B., Hu, W., Liu, Y., Yu, G.: Synthesis and morphology transformation of single-crystal graphene domains based on activated carbon dioxide by chemical vapor deposition. J. Mater. Chem. C **1**(17), 2990–2995 (2013). https://doi.org/10.1039/c3tc30124a
46. Lv, H., Pan, Q., Song, Y., Liu, X.X., Liu, T.: A review on nano-/microstructured materials constructed by electrochemical technologies for supercapacitors. Nano-Micro Lett. **12**(1), (2020). https://doi.org/10.1007/s40820-020-00451-z. (Springer Singapore)
47. Lv, W., Tang, D.M., He, Y.B., You, C.H., Shi, Z.Q., Chen, X.C., Chen, C.M., Hou, P.X., Liu, C., Yang, Q.H.: Low-temperature exfoliated graphenes: vacuum-promoted exfoliation and electrochemical energy storage. ACS Nano **3**(11), 3730–3736 (2009). https://doi.org/10.1021/nn900933u
48. Mattevi, C., Kim, H., Chhowalla, M.: A review of chemical vapour deposition of graphene on copper. J. Mater. Chem. **21**(10), 3324–3334 (2011). https://doi.org/10.1039/c0jm02126a
49. Merlen, A., Buijnsters, J.G., Pardanaud, C.: A guide to and review of the use of multiwavelength Raman spectroscopy for characterizing defective aromatic carbon solids: from graphene to amorphous carbons. Coatings **7**(10), (2017). https://doi.org/10.3390/coatings7100153
50. Meyer, J.C., Geim, A.K., Katsnelson, M.I., Novoselov, K.S., Booth, T.J., Roth, S.: The structure of suspended graphene sheets. Nature **446**(7131), 60–63 (2007). https://doi.org/10.1038/nature05545
51. Meyer, J.C., Kisielowski, C., Erni, R., Rossell, M.D., Crommie, M.F., Zettl, A.: Direct imaging of lattice atoms and topological defects in graphene membranes. Nano Lett. **8**(11), 3582–3586 (2008). https://doi.org/10.1021/nl801386m
52. Miller, J.R., Outlaw, R.A., Holloway, B.C.: Graphene double-layer capacitor with ac line-filtering performance. Science **329**(5999), 1637–1639 (2010). https://doi.org/10.1126/science.1194372

53. Ni, Z.H., Chen, W., Fan, X.F., Kuo, J.L., Yu, T., Wee, A.T.S., Shen, Z.X.: Raman spectroscopy of epitaxial graphene on a SiC substrate. Phys. Rev. B-Condens. Matter Mater. Phys. **77**(11), 1–6 (2008). https://doi.org/10.1103/PhysRevB.77.115416
54. Ni, Z., Wang, Y., Yu, T., Shen, Z.: Raman spectroscopy and imaging of graphene. Nano Res. **1**(4), 273–291 (2008). https://doi.org/10.1007/s12274-008-8036-1
55. Ortiz Balbuena, J., Tutor De Ureta, P., Rivera Ruiz, E., Mellor Pita, S.: Enfermedad de Vogt-Koyanagi-Harada. Med. Clin. **146**(2), 93–94 (2016). https://doi.org/10.1016/j.medcli.2015.04.005
56. Othman, M., Ritikos, R., Muhammad Hafiz, S., Khanis, N.H., Abdul Rashid, N.M., Abdul Rahman, S.: Low-temperature plasma-enhanced chemical vapour deposition of transfer-free graphene thin films. Mater. Lett. **158**, 436–438 (2015). https://doi.org/10.1016/j.matlet.2015.06.039
57. Panchakarla, L.S., Govindaraj, A., Rao, C.N.R.: Boron- and nitrogen-doped carbon nanotubes and graphene. Inorg. Chim. Acta **363**(15), 4163–4174 (2010). https://doi.org/10.1016/j.ica.2010.07.057
58. Qian, W., Liu, T., Wang, Z., Yu, H., Li, Z., Wei, F., Luo, G.: Effect of adding nickel to iron-alumina catalysts on the morphology of as-grown carbon nanotubes. Carbon **41**(13), 2487–2493 (2003). https://doi.org/10.1016/S0008-6223(03)00324-5
59. Qin, B., Zhang, T., Chen, H., Ma, Y.: The growth mechanism of few-layer graphene in the arc discharge process. Carbon **102**, 494–498 (2016). https://doi.org/10.1016/j.carbon.2016.02.074
60. Rasool, H.I., Song, E.B., Mecklenburg, M., Regan, B.C., Wang, K.L., Weiller, B.H., Gimzewski, J.K.: Atomic-scale characterization of graphene grown on copper (100) single crystals. J. Am. Chem. Soc. **133**(32), 12536–12543 (2011). https://doi.org/10.1021/ja200245p
61. Reina, A., Jia, X., Ho, J., Nezich, D., Son, H., Bulovic, V., Dresselhaus, M.S., Kong, J.: Large Area, Few-Layer Graphene Films on Arbitrary Substrates by Chemical Vapor Deposition, pp. 1–6 (2008)
62. Soldano, C., Mahmood, A., Dujardin, E.: Production, properties and potential of graphene. Carbon **48**(8), 2127–2150 (2010). https://doi.org/10.1016/j.carbon.2010.01.058
63. Stankovich, S., Dikin, D.A., Piner, R.D., Kohlhaas, K.A., Kleinhammes, A., Jia, Y., Wu, Y., Nguyen, S.B.T., Ruoff, R.S.: Synthesis of graphene-based nanosheets via chemical reduction of exfoliated graphite oxide. Carbon **45**(7), 1558–1565 (2007). https://doi.org/10.1016/j.carbon.2007.02.034
64. Su, Y., Zhang, Y.: Carbon nanomaterials synthesized by arc discharge hot plasma. Carbon **83**(November), 90–99 (2015). https://doi.org/10.1016/j.carbon.2014.11.023
65. Subrahmanyam, K.S., Panchakarla, L.S., Govindaraj, A., Rao, C.N.R.: Simple method of preparing graphene flakes by an arc-discharge method. J. Phys. Chem. C **113**(11), 4257–4259 (2009). https://doi.org/10.1021/jp900791y
66. Tan, H., Wang, D., Guo, Y.: A strategy to synthesize multilayer graphene in arc-discharge plasma in a semi-opened environment. Materials **12**(14), (2019). https://doi.org/10.3390/ma12142279
67. Tang, T., Chen, X., Meng, X., Chen, H., Ding, Y.: Synthesis of multiwalled carbon nanotubes by catalytic combustion of polypropylene. Angew. Chem.-Int. Ed. **44**(10), 1517–1520 (2005). https://doi.org/10.1002/anie.200461506
68. Teng, C., Xie, D., Wang, J., Yang, Z., Ren, G., Zhu, Y.: Ultrahigh conductive graphene paper based on ball-milling exfoliated graphene. Adv. Funct. Mater. **27**(20), (2017). https://doi.org/10.1002/adfm.201700240
69. Tontegode, A.Y.: Carbon on transition metal surfaces. Prog. Surf. Sci. **38**(3–4), 201–429 (1991). https://doi.org/10.1016/0079-6816(91)90002-L
70. Tu, Z., Liu, Z., Li, Y., Yang, F., Zhang, L., Zhao, Z., Xu, C., Wu, S., Liu, H., Yang, H., Richard, P.: Controllable growth of 1–7 layers of graphene by chemical vapour deposition. Carbon **73**, 252–258 (2014). https://doi.org/10.1016/j.carbon.2014.02.061
71. Tung, V.C., Allen, M.J., Yang, Y., Kaner, R.B.: High-throughput solution processing of large-scale graphene. Nat. Nanotechnol. **4**(1), 25–29 (2009). https://doi.org/10.1038/nnano.2008.329

72. Viera Skakalova, A.B.K.: Graphene Properties, Preparation, Characterisation and Devices (2014)
73. Wang, G., Wang, B., Park, J., Wang, Y., Sun, B., Yao, J.: Highly efficient and large-scale synthesis of graphene by electrolytic exfoliation. Carbon **47**(14), 3242–3246 (2009). https://doi.org/10.1016/j.carbon.2009.07.040
74. Wang, J.J., Zhu, M.Y., Outlaw, R.A., Zhao, X., Manos, D.M., Holloway, B.C., Mammana, V.P.: Free-standing subnanometer qraphite sheets. Appl. Phys. Lett. **85**(7), 1265–1267 (2004). https://doi.org/10.1063/1.1782253
75. Wang, J., Manga, K.K., Bao, Q., Loh, K.P.: High-yield synthesis of few-layer graphene flakes through electrolyte. J. Am. Chem. Soc **133**, 8888–8891 (2011)
76. Wang, Z., Li, N., Shi, Z., Gu, Z.: Low-cost and large-scale synthesis of graphene nanosheets by arc discharge in air. Nanotechnology **21**(17), (2010). https://doi.org/10.1088/0957-4484/21/17/175602
77. Wintterlin, J., Bocquet, M.L.: Graphene on metal surfaces. Surf. Sci. **603**(10–12), 1841–1852 (2009). https://doi.org/10.1016/j.susc.2008.08.037
78. Wu, T., Ding, G., Shen, H., Wang, H., Sun, L., Jiang, D., Xie, X., Jiang, M.: Triggering the continuous growth of graphene toward millimeter-sized grains. Adv. Func. Mater. **23**(2), 198–203 (2013). https://doi.org/10.1002/adfm.201201577
79. Wu, X., Liu, Y., Yang, H., Shi, Z.: Large-scale synthesis of high-quality graphene sheets by an improved alternating current arc-discharge method. RSC Adv. **6**(95), 93119–93124 (2016). https://doi.org/10.1039/c6ra22273k
80. Wu, Y., Wang, B., Ma, Y., Huang, Y., Li, N., Zhang, F., Chen, Y.: Efficient and large-scale synthesis of few-layered graphene using an arc-discharge method and conductivity studies of the resulting films. Nano Res. **3**(9), 661–669 (2010). https://doi.org/10.1007/s12274-010-0027-3
81. Xu, X., Zhang, Z., Qiu, L., Zhuang, J., Zhang, L., Wang, H., Liao, C., Song, H., Qiao, R., Gao, P., Hu, Z., Liao, L., Liao, Z., Yu, D., Wang, E., Ding, F., Peng, H., Liu, K.: Ultrafast growth of single-crystal graphene assisted by a continuous oxygen supply. Nat. Nanotechnol. **11**(11), 930–935 (2016). https://doi.org/10.1038/nnano.2016.132
82. Xu, Z., Li, H., Li, W., Cao, G., Zhang, Q., Li, K., Fu, Q., Wang, J.: Large-scale production of graphene by microwave synthesis and rapid cooling. Chem. Commun. **47**(4), 1166–1168 (2011). https://doi.org/10.1039/c0cc03520c
83. Yazdi, G.R., Vasiliauskas, R., Iakimov, T., Zakharov, A., Syväjärvi, M., Yakimova, R.: Growth of large area monolayer graphene on 3C-SiC and a comparison with other SiC polytypes. Carbon **57**, 477–484 (2013). https://doi.org/10.1016/j.carbon.2013.02.022
84. Yousef, S., Khattab, A., Osman, T.A., Zaki, M.: Effects of increasing electrodes on CNTs yield synthesized by using arc-discharge technique. J. Nanomater. **2013**, (2013). https://doi.org/10.1155/2013/392126
85. Yu, Q., Jauregui, L.A., Wu, W., Colby, R., Tian, J., Su, Z., Cao, H., Liu, Z., Pandey, D., Wei, D., Chung, T.F., Peng, P., Guisinger, N.P., Stach, E.A., Bao, J., Pei, S.S., Chen, Y.P.: Control and characterization of individual grains and grain boundaries in graphene grown by chemical vapour deposition. Nat. Mater. **10**(6), 443–449 (2011). https://doi.org/10.1038/nmat3010
86. Yuan, C., Hsin, Y., Jong, L.: Growth of large-sized graphene thin-films by liquid precursor-based chemical vapor deposition under atmospheric pressure (2011)
87. Zhang, W., Cui, J., Tao, C., Wu, Y., Li, Z., Ma, L., Wen, Y., Li, G.: A strategy for producing pure single-layer graphene sheets based on a confined self-assembly approach. Angew. Chem. **121**(32), 5978–5982 (2009). https://doi.org/10.1002/ange.200902365
88. Zhu, M., Wang, J., Holloway, B.C., Outlaw, R.A., Zhao, X., Hou, K., Shutthanandan, V., Manos, D.M.: A mechanism for carbon nanosheet formation. Carbon **45**(11), 2229–2234 (2007). https://doi.org/10.1016/j.carbon.2007.06.017

Graphene Oxide

A. Kingson Solomon Jeevaraj and M. Muthuvinayagam

Abstract Graphene is a two-dimensional, crystalline allotrope with a hexagonal lattice structure made from pure carbon atoms. Graphene oxide (GO) is a single monomolecular layer of graphite containing epoxide, carbonyl, carboxyl, and hydroxyl groups. The GO film can be made semi-metallic, insulating or semiconductive and still maintain optical transparency. An overview of graphene oxide's structure, properties, and fabrication is provided in this article. It discusses in detail the methods of characterizing graphene oxide, including transmission electron microscopy, X-ray diffraction analysis, Fourier transform infrared spectroscopy, thermogravimetry, elemental analysis, and X-ray photoelectron spectroscopy (XPS), and the applications of graphene oxide.

Keywords Graphene oxide · Properties · Preparation · Characterization · Applications

1 Introduction

The graphene material is made up of carbon atoms that are arranged in a hexagonal pattern. Due to being the thinnest known material today, graphene is considered to be two-dimensional. Aside from being the strongest material on earth, graphene is also considered to be one of the most electrically and thermally conductive material. A wide range of applications can be found for graphene, including electronics, medicine, aviation, and many others.

As graphene is expensive and difficult to fabricate, and great efforts have been significantly devoted to find ways to make and use graphene derivatives that are both effective and inexpensive. For instance, GO, which is a material that is formed

A. K. S. Jeevaraj (✉)
Department of Physics, LRG Government Arts College for Women, Tirupur, Tamilnadu, India
e-mail: drkingson@gmail.com

M. Muthuvinayagam (✉)
Department of Physical Sciences, Saveetha School of Engineering, Saveetha University (SIMATS), Chennai, Tamil Nadu 600 077, India
e-mail: muthuvinayaggamm.sse@saveetha.com

© The Author(s), under exclusive license to Springer Nature Singapore Pte Ltd. 2023

R. T. Subramaniam et al. (eds.), *Graphene*, Engineering Materials,
https://doi.org/10.1007/978-981-99-1206-3_5

by the powerful oxidation of inexpensive and abundant graphite. Moreover, GO is a monomolecular layer of graphite with a variety of oxygen-containing functional groups, including epoxide, carbonyl, carboxyl, and hydroxyl groups [1]. Furthermore, GO is dispersible in water (as well as other solvents) and can even be used to make graphene. Despite knowing that GO has a lower conductivity compared to graphene, GO, however, there are numerous processes that exist to enhance its properties. For instance, powders, dispersions, and coatings are all the common forms.

2 Structure and Properties

Graphene is a two-dimensional, crystalline allotrope with a hexagonal lattice structure made from pure carbon atoms. They are best known for its unique properties containing high optical transparency, the best heat conductivity at room temperature and the ability to be flexible all within a strong, nano-sized material (Fig. 1).

GO has significant electronic, thermal and physical properties which makes it an excellent product for use in not only research and the development of advanced materials, but also in revolutionizing the energy storage industry as well.

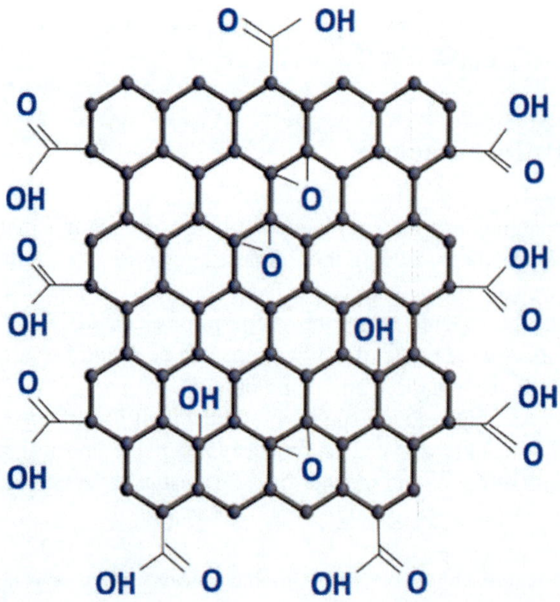

Fig. 1 The graphene oxide molecular structure consists of carbon, hydrogen and oxygen. (https://www.acsmaterial.com/blog-detail/graphene-oxide.html, free access on 15-01-2023)

2.1 Thermal and Electronic Properties

GO usually decomposes when heated due to the release of water absorbed. However, when oxygen-containing groups are removed, the thermal stability gradually increases [2]. Deposition and reduction parameters may also be adjusted to tune the electronic properties of graphene oxide. A GO film can be made semi-metallic, insulating or semiconductive, while preserving optical transparency [2].

2.2 Optical Properties

In general, GO exhibits a high optical transmission in the visible spectrum. However, this optical transmittance may be adjusted based on the film thickness and the extent of reduction. The wavelength of GO was found to be at 550 nm particularly when fabricated from 0.5 mg/ml suspensions, and the optical transmittance was at approximately 96%. Alternatively, when the wavelength was further reduced (<30 nm), a semi-transparent material resulted, while an atomically thin GO will be highly transparent [2].

2.3 Mechanical Properties

The mechanical properties of GO films can be generally tuned either by doping or compositing. GO films that have thickness in the range of mm are usually acquired through filtration or slow evaporation of graphene oxide solution [2] (Table 1).

Table 1 Mechanical properties of unmodified and modified GO papers [3]

Materials	Maximum Young's modulus, EE (GPa)*	Fracture strain, ε (%)	Intrinsic strength, σ (MPa)
Unmodified paper	25.6 ± 1.1	0.40 ± 0.03	81.9 ± 5.3
rinsed Ca-modified paper	28.1 ± 1.2	0.50 ± 0.06	125.8 ± 13.6
As-prepared Ca-modified paper	21.5 ± 1.5	0.41 ± 0.15	75.4 ± 21.6
Rinsed Mg-modified paper	27.9 ± 1.8	0.33 ± 0.08	80.6 ± 16.5
Polyallylamine (PAA)-modified paper	33.3 ± 2.7	0.32 ± 0.08	91.9 ± 22.4
As-prepared Mg-modified paper	24.6 ± 1.4	0.40 ± 0.04	87.9 ± 14.2

* EE is the maximum Young's modulus in the linear region ($\sigma = 10$ MPa)

3 Fabrication of GO

3.1 *Brodie Method*

The first documented synthesis of graphitic oxide materials was attributed by Brodie in 1859 [4]. The researcher's work was focused on finding the weight of the graphite and during that time, a series of chemical reactions were investigated to elucidate the properties of this novel material, which was GO. Thus, graphite was mixed with potassium chlorate ($KClO_3$) and solubilized in fumic nitric acid to oxidize the sample. Further oxidation processes were carried out on the sample till any changes were visible. The elemental analysis revealed a composition of circa 60% C, 2% H and 38% O. The final product resulted in a mixture of graphene and graphite oxide, which is soluble in pure water.

3.2 *Staudenmaier Method*

The Brodie's reaction was improved in 1898 by Staudenmaier [5] by adding sulfuric acid ($H_2\ SO_4$) to increase the acidity of the mixture and several aliquots of solid $KClO_3$ over the course of the reaction. In spite of this, the Brodie and Staudenmaier's method generated toxic gas (ClO_2) that rapidly decomposes in air, causing explosions. As a result of these modifications, the graphitic material became more oxidized and the reaction was simplified.

3.3 *Hummer's Method*

GO was synthesized by Hummer's method [6] through the oxidation of graphite. The following are the steps involved for the fabrication of GO via Hummer's method: Firstly, two grams of graphite flakes and two grams of $NaNO_3$ were blended in 50 mL of $H_2\ SO_4$ (98%) in a 1000 mL volumetric flask and heated with continuous stirring for two hours. The reaction temperature was kept below 15 °C while 6 g of potassium permanganate ($KMnO_4$) was slowly added to the suspension. After removing the ice bath, the mixture was stirred for 48 h at 35 °C until it became pasty brownish in colour. Upon diluting the mixture with 100 ml of water, the reaction temperature quickly increased to 98 °C with effervescence, resulting in brown coloration. In addition, 200 ml of water was added to the mixture and stirred continuously. A final treatment of 10 ml hydrogen peroxide (H_2O_2) transformed the solution into yellow colour. In order to purify the mixture, it was rinsed and centrifuged several times with 10% HCl, followed by several rinses with deionized water (DI). Finally, the GO powder was obtained by filtering and drying under vacuum at room temperature.

3.4 Modified Hummer's Method

This modified method involves both oxidation and exfoliation of graphite sheets due to thermal treatment of solution. The following are the steps involved to fabricate GO via the modified Hummer's method [7]:

Firstly, graphite flakes (2 g) and $NaNO_3$ (2 g) were mixed in 90 mL of H_2SO_4 (98%) in a 1000 ml volumetric flask kept under at ice bath (0–5 °C) with continuous stirring for 4 h.

Potassium permanganate (12 g) was added to the suspension very slowly and the reaction temperature was maintained lower than 15 °C. Dilute the mixture with very slow addition of 184 ml water and kept under stirring for 2 h. After removing the ice bath, the mixture was stirred at 35 °C for 2 h. The above mixture was kept in a reflux system at 98 °C for 10–15 min. After 10 min, the change in the temperature to 30 °C resulted to a brown coloured solution. Subsequently, the temperature was then reduced to 25 °C and maintained for 2 h. Finally, the solution was treated with 40 ml H_2O_2, resulting in the colour to change to bright yellow.

Furthermore, 200 ml of water was added in two separate beakers and equal amount of solution was prepared and added followed by stirring for 1 h. The solution was then kept without stirring for 3–4 h, in order to allow the particles to settle at the bottom and the remaining water was then filtered out. The resulting mixture was washed repeatedly by centrifugation with 10% HCl and then with DI water several times until a gel like substance was formed (pH- neutral). After centrifugation, the gel like substance was vacuum dried at 60 °C for more than 6 h, resulting in the final product, GO powder.

3.5 Tour Method

In 2010, at Rice University [8], an improved version of Hummers method was proposed by Tour's group. They have replaced the sodium nitride ($NaNO_3$) with ortho-phosphoric acid (H_3PO_4) in a mixture of H_2SO_4/H_3PO_4 (9:1) and increased the amount of $KMnO_4$. Here, there was no generation of toxic gases, such as NO_2, N_2O_4 or ClO_2, and this was considered as the advantage of this method and also, the temperature was easily controllable in the Tour Method. In addition, the authors claimed that the presence of ortho- phosphoric acid generated a more intact graphitic basal plane.

4 Characterization of GO

4.1 Morphology

(i) Transmission Electron Microscopy (TEM)

TEM are microscopes that use a particle beam of electrons to visualize specimens and generate a highly-magnified image. TEM can magnify objects up to 2 million times. In order to get a better idea of just how small that is, think of how small a cell is. In other words, TEM has become so valuable within the biological and medical fields.

TEM employs a high voltage electron beam in order to create an image. An electron gun at the top of a TEM emits electrons that travel through the microscope's vacuum tube. Rather than having a glass lens focusing the light (as in the case of light microscopes), the TEM employs an electromagnetic lens which focuses the electrons into a very fine beam. This beam then passes through the specimen, which is very thin, and the electrons either scatter or hit a fluorescent screen at the bottom of the microscope. An image of the specimen with its assorted parts shown in different shades according to its density appears on the screen. This image can be then studied directly within the TEM or photographed. Figure 2 shows a diagram of TEM and its basic parts.

(ii) Dynamic Force Microscopy

Dynamic Force Mode is also known as tapping mode, intermittent-contact mode, acoustic AC mode or oscillating mode respectively. Figure 3 shows the schematic of dynamic mode operation. A stiff cantilever is oscillated closer to the sample than in the non-contact mode. Part of the oscillation extends into the repulsive regime, so the tip intermittently touches or taps the surface. Very stiff cantilevers are typically used, as tips can get stuck in the water contamination layer. The advantage of tapping the surface improved the lateral resolution on soft samples. Lateral forces such as drag, common in the contact mode, are virtually eliminated. For poorly adsorbed specimens on a substrate surface, the advantage is clearly seen.

4.2 Structure of GO Sheet

XRD Analysis

XRD analysis is used to determine the average crystalline properties of the GO sheet. X-ray diffraction is a powerful non-destructive technique for characterizing crystalline materials. It provides information on structures, phases, preferred crystal orientations (texture), and other structural parameters, such as average grain size, crystallinity, strain, and crystal defects. XRD peaks are produced by constructive

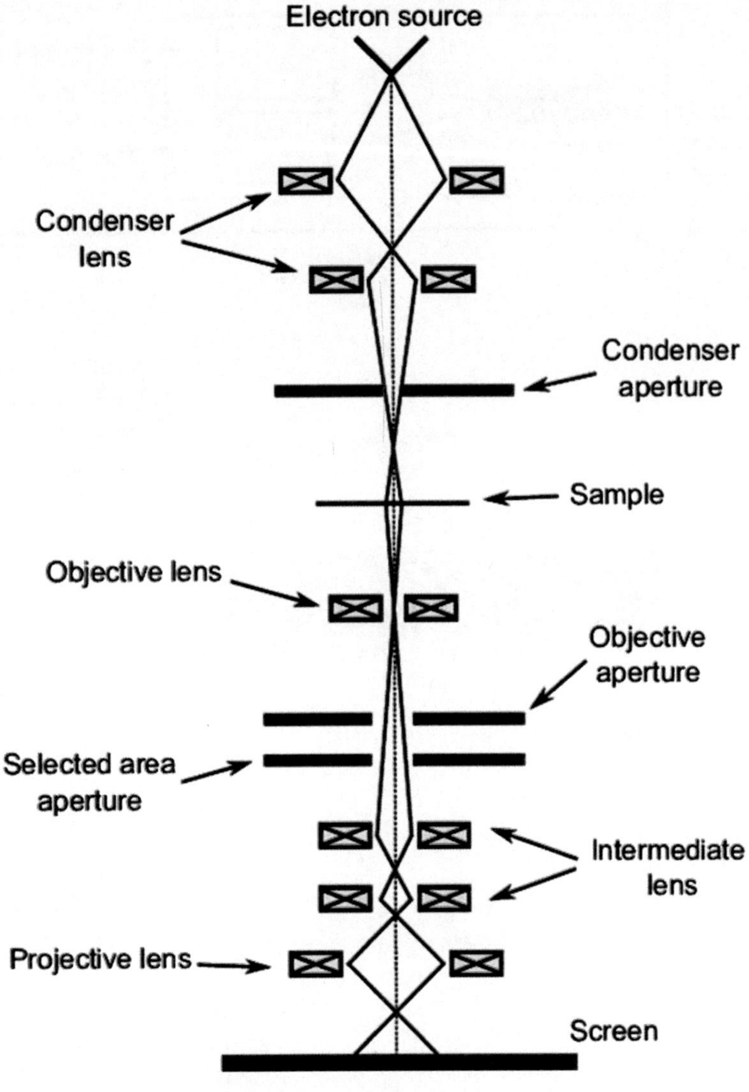

Fig. 2 Schematic diagram of TEM and its basic parts

interference of a monochromatic beam of X-rays scattered at specific angles from each set of lattice planes in a sample. The peak intensities are determined by the atomic positions within the lattice planes. Consequently, the XRD pattern is the fingerprint of periodic atomic arrangements in a given material. An online search of a standard database for X-ray powder diffraction patterns enables quick phase identification for a large variety of crystalline samples. Figure 4 shows the analysis of x-ray diffraction.

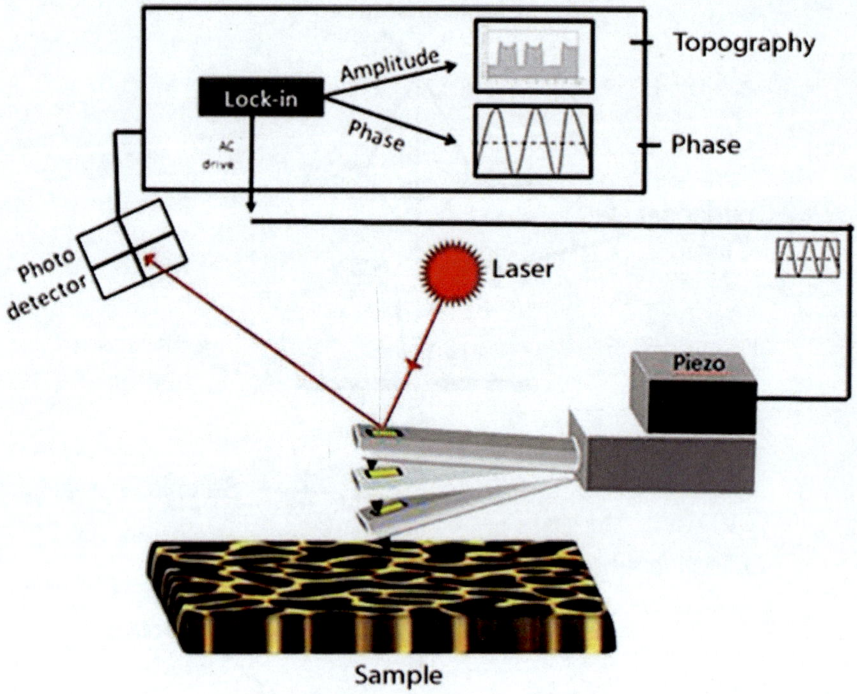

Fig. 3 Schematic diagram of dynamic mode operation

Fig. 4 Illustration of X- ray diffraction analysis

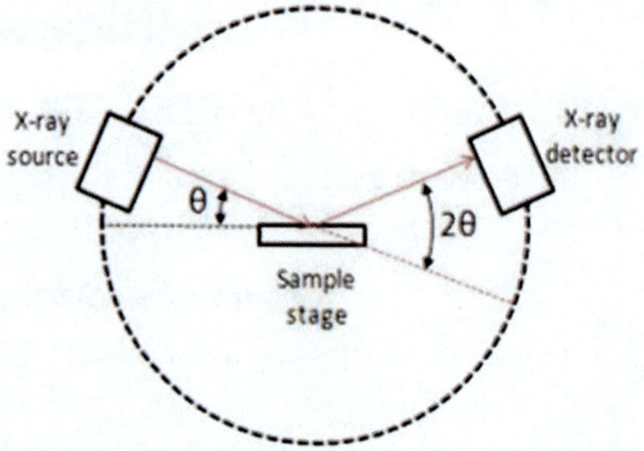

4.3 Fourier Transform Infrared Spectroscopy (FTIR) Analysis

FTIR Analysis measures the infrared region of the electromagnetic radiation spectrum, which has a longer wavelength and a lower frequency than visible light, and is measurable in a sample when submitted to infrared radiation (IR). The basic theory at work is that the bonds between different elements absorb light at different frequencies. The light is measured using an infrared spectrometer which produces the output of an infrared spectrum. The IR spectrum is a graph of infrared light absorbance by the substance on the vertical axis and the frequency (wavelength) on the horizontal axis.

FTIR analysis measures the ranges of wavelengths in the infrared region that are absorbed by a material. This is accomplished through the application of infrared radiation (IR) to samples of a material. The sample's absorbance of the infrared light's energy at various wavelengths is measured to determine the material's molecular composition and structure.

Unknown materials are identified by searching the spectrum against a database of reference spectra. Materials can be quantified using the FTIR materials characterization technique as long as a standard curve of known concentrations of the component of interest can be created.

FTIR Analysis can be used to identify unknown materials, additives within polymers, surface contamination on a material, and more. The results of the tests can pinpoint a sample's molecular composition and structure.

A simple device called an interferometer is used to identify samples by producing an optical signal with all the IR frequencies encoded into it. The signal can be measured quickly.

Then, the signal is decoded by applying a mathematical technique known as Fourier transformation. This computer-generated process then produces a mapping of the spectral information. The resulting graph is the spectrum which is then searched against reference libraries for identification. Figure 5 shows the instrumentation of FTIR analysis.

With the microscope attachment, samples as small as 20 μm can be analyzed. This allows quick and cost effective identification of unknown particles, residues, films or fibers. FTIR testing can also measure levels of oxidation in some polymers or degrees of cure in other polymers as well as quantifying contaminants or additives in materials.

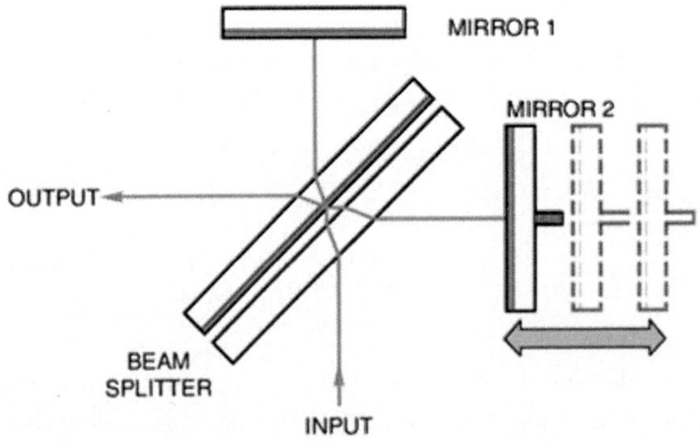

Fig. 5 Instrumentation of FTIR analysis

4.4 Optical Absorption Analysis

The optical phenomenon known as diffuse reflectance [9] is commonly used in the UV-visible, near-infrared (NIR), and mid-infrared (sometimes called DRIFT or DRIFTS) regions to obtain molecular spectroscopic information. It is usually used to obtain spectra of powders with minimum sample preparation. A reflectance spectrum is obtained by the collection and analysis of surface-reflected electromagnetic radiation as a function of frequency (v, cm^{-1}) or wavelength (A, nm). Two different types of reflection can occur: regular or specular reflection usually associated with reflection from smooth, polished surfaces like mirrors, and diffuse reflection associated with reflection from the so-called mat or dull surfaces textured like powders. Techniques such as external reflectance and total internal reflectance spectroscopy use the phenomenon of specular reflection to obtain spectroscopic information. In diffuse reflectance spectroscopy, electromagnetic radiation reflected from the dull surfaces is collected and analysed. If a sample to be analysed is not shiny, and for whatever reason is not amenable to conventional transmission spectroscopy, diffuse reflectance spectroscopy is a logical alternative.

4.5 Thermal Stability of GO Sheet

Thermogravimetric analysis (TGA) is conducted to test the thermal stability of the GO sheet.

TGA is a technique in which, upon heating a material, its weight increases or decreases. A TGA consists of a sample pan that is supported by a precision balance. That pan resides in a furnace and is heated or cooled during the experiment. The mass of the sample is monitored during the experiment. A sample purge gas controls the sample environment. This gas may be inert or a reactive gas that flows over the sample and exits through an exhaust.

These instruments can quantify the loss of water, loss of solvent, loss of plasticizer, decarboxylation, pyrolysis, oxidation, decomposition, weight % filler, amount of metallic catalytic residue remaining on carbon nanotubes, and weight % ash. All these quantifiable applications are usually done upon heating, but there are some experiments where information may be obtained upon cooling. Figure 6 shows the schematic of thermogravimetric analysis.

4.6 Elemental Analysis of GO Sheet

The elemental analysers operate by heating an element quickly to a sufficiently high temperature so that it combusts. It is then passed through the analyser in its

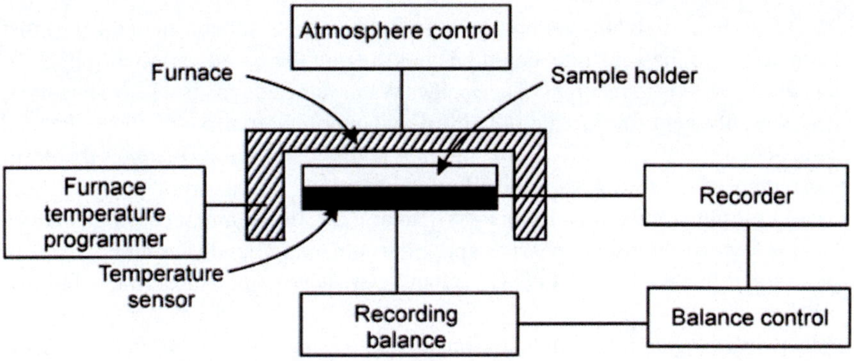

Fig. 6 Schematic of Thermogravimetric analysis

gaseous state. A detector detects the elements present, and the researcher reads this information on a computer screen.

4.7 The XPS Analysis of GO Sheet

X-ray photoelectron spectroscopy (XPS), also known as electron spectroscopy for chemical analysis (ESCA), is a technique employed to analyse a material's surface chemistry. XPS can measure the elemental composition as well as the chemical and electronic state of the atoms within a material.

XPS spectra are obtained by irradiating a solid surface with a beam of X-rays and measuring the kinetic energy of electrons that are emitted from the top (1–10 nm) of the material. A photoelectron spectrum is recorded by counting ejected electrons over a range of kinetic energies. The energies and intensities of the photoelectron peaks enable identification and quantification of all surface elements (except hydrogen).

5 Applications of GO

GO can be used in various applications, including the following [3, 10–12]:

- **Conductive Transparent Coatings**—Transparent protective and conductive coatings are one of the first and widely discussed applications for reduced GO (rGO) [13–16]. Films with thickness of several atomic layers could be easily deposited on any hydrophilic surface by simple drop-casting method or spin coating [17] and subsequent evaporating water. In the case of using electrically conductive substrate the smoothness and density of the film can be enhanced by using electrophoresis [18]. If the surface of substrate is hydrophobic, then the dispersion in suitable organic solvent can be used. The reduction of GO films to rGO must be carried out avoiding damage of substrate. A suitable method, the most compatible to the substrate, could be chosen from the broad variety of physical, chemical, or hybrid methods. The van der Waals adhesion to the substrate in most cases is sufficient for durable rGO films and coatings. It also could be enhanced using chemical bonding to carboxyl groups at the edges of rGO platelets that form a structure of the film. Using GO allows fabrication of transparent graphene electrodes of unrestricted area for displays, solar cells and luminescent light sources [19]. Graphene is less expensive and environmental friendly in comparison to traditional transparent conductor ITO and can successfully substitute it in most applications.

- **Electronics**—examples include transparent conductors, field-effect devices, flexible electronic materials, and electrical sensors.
- **Water purification**—GO membranes have high water permeability and stability making them a promising material for water treatment application.
- **Optics**—tuneable optical properties of GO have led to promising applications in the field of fluorescence quenching, surface-enhanced Raman scattering, optical sensing/detecting and nonlinear optical materials.
- **Optoelectronics**—examples include photovoltaic devices and light-emitting devices. Large area of rGO transparent electrodes can become indispensable component of low-cost flexible solar cells based on organic materials. Recently, GO and chemically modified rGO were used as components for electron transport layers in new intensively developing organic perovskite based solar cells [20]. In short, GO or rGO are also useful materials for counter electrode in dye-sensitized ultralow cost solar cells based on the liquid electrolyte [21].
- **Electrode material**—GO's have a high surface area that makes them suitable for use as the electrode material for capacitors, solar cells and batteries. **Supercapacitors**—Although supercapacitors fall behind lithium batteries in energy capacity, however, they are able to provide a much greater peak power. It is important for a variety of applications in industry, transport and communications. Supercapacitors store energy in electric double layer capacitors (EDLC) [22], formed by ions of liquid, solid or ionic liquid electrolyte accumulated on the surface of the electrode material. Therefore, the electrode material must be chemically inert, have the largest possible specific area and opened porous structure. It also must have smallest resistance and mechanical durability to provide maximum number of charge–discharge cycles at highest value of current.
- **Enhancer**—GO can be easily combined with other composite materials to improve their properties such as conductivity, elasticity, and tensile strength.
- **Bio-based applications**—examples include electrochemical biosensors and optical biosensors, where GOs are used to detect DNA, enzymes and other biomolecules with high selectivity and sensitivity. Graphene oxide may also be used in disease detection through its fluorescent property.

References

1. Ray, S.C.: Chapter 2-application and uses of graphene oxide and reduced graphene oxide. In: Ray, S.C. (ed.) Micro and Nano Technologies, Applications of Graphene and Graphene-Oxide Based Nanomaterials, pp. 39–55. William Andrew Publishing (2015). ISBN 9780323375214. https://doi.org/10.1016/B978-0-323-37521-4.00002-9
2. Paulchamy, B., Arthl, G., Lignesh, B.D.: A simple approach to stepwise synthesis of graphene oxide nanomaterial. J. Nanomed. Nanotechnol. **6**, 1 (2015)
3. Zhao, J., Liu, L., Li, F.: Graphene Oxide: Physics and Applications. Springer (2015)
4. Brodie, B.C.: On the atomic weight of graphite. Philos. Trans. R. Soc. Lond. **149**, 249–259 (1859)
5. Staudenmaier, L.: Verfahren zur Darstellung der Graphitsäure. Ber. Dtsch. Chem. Ges. **31**(2), 1481–1487 (1898)
6. Hummers, W.S., Offeman, R.E.: Preparation of graphitic oxide. J. Am. Chem. Soc. **80**(6), 1339–1339 (1958)
7. Song, J., Wang, X., Chang, C.T.: Preparation and characterization of graphene oxide. J. Nanomater. **2014**(276143), 6 (2014). https://doi.org/10.1155/2014/276143
8. Marcano, D.C., Kosynkin, D.V., Berlin, J.M., Sinitskii, A., Sun, Z., Slesarev, A., Alemany, L.B., Lu, W.: Tour JM improved synthesis of graphene oxide. ACS Nano **4**(8), 4806–4814 (2010)
9. Mirabella, F.M. (ed.): Modern Techniques in Applied Molecular Spectroscopy. Techniques in Analytical Chemistry Series. © Wiley & Sons, Inc. (1998). ISBN 0-471-12359-5
10. Graphene Oxide: Introduction and Market News (2019). https://www.graphene-info.com/graphene-oxide. Last accessed 04 Dec 2019
11. Smith, A.T., LaChance, A.M., Zeng, S., Liu, B., Sun, L.: Synthesis, properties, and applications of graphene oxide/reduced graphene oxide and their nanocomposite. Nano Mater. Sci. **1**, 31–47 (2019)
12. Dideikin, A.T., Vul, A.Y.: Graphene oxide and derivatives: the place in graphene family. Front. Phys., (2019). https://doi.org/10.3389/fphy.2018.00149
13. Novoselov, K.S., Fal'ko, V.I., Colombo, L., Gellert, P.R., Schwab, M.G., Kim, K.: A roadmap for graphene. Nature **490**, 192–200 (2012). https://doi.org/10.1038/nature11458
14. Wu, S.X., Yin, Z.Y., He, Q.Y., Huang, X., Zhou, X.Z., Zhang, H.: Electrochemical deposition of semiconductor oxides on reduced graphene oxide-based flexible, transparent and conductive electrodes. J. Phys. Chem. C **114**, 11816 (2010). https://doi.org/10.1021/jp103696u
15. Eda, G., Fanchini, G., Chhowalla, M.: Large-area ultrathin films of reduced graphene oxide as a transparent and flexible electronic material. Nat Nanotech. **3**, 270–274 (2008). https://doi.org/10.1038/nnano.2008.83
16. Zhang, X., Yan, X., Chen, J., Zhao, J.: Large-size graphene microsheets as a protective layer for transparent conductive silver nanowire film heaters. Carbon **69**, 437–443 (2014). https://doi.org/10.1016/j.carbon.2013.12.046
17. Yamaguchi, H., Eda, G., Mattevi, C., Kim, H.K., Chhowalla, M.: Highly uniform 300 mm wafer-scale deposition of single and multilayered chemically derived graphene thin films. ACS Nano **4**, 524–528 (2010). https://doi.org/10.1021/nn901496p
18. Lia, S., Zhang, Q., Lu, Y., Ji, D., Zhang, D., Wu, J., et al.: One step electrochemical deposition and reduction of graphene oxide on screen printed electrodes for impedance detection of glucose. Sens. Actuators B **244**, 290–298 (2017). https://doi.org/10.1016/j.snb.2016.12.142
19. Morales-Masis, M., De Wolf, S., Woods-Robinson, R., Ager, J.W., Ballif, C.: Transparent electrodes for efficient optoelectronics. Adv. Electron. Mater. **5**, 1600529 (2017). https://doi.org/10.1002/aelm.201600529
20. Assadi, M.K., Bakhoda, S., Saidur, R., Hanaei, H.: Recent progress in perovskite solar cells. Renew. Sustain. Ener Rev. **81**, 2812–2822 (2018). https://doi.org/10.1016/j.rser.2017.06.088

21. Williams, G., Seger, B., Kamat, P.V.: TiO_2-graphene nanocomposites. UV-assisted photocatalytic reduction of graphene oxide. ACS Nano **2**, 1487–1491 (2008). https://doi.org/10.1021/nn800251f
22. Wang, F., Wu, X., Yuan, X., Liu, Z., Zhang, Y., Fu, L., et al.: Latest advances in supercapacitors: from new electrode materials to novel device designs. Chem. Soc. Rev. **46**, 6816–6854 (2017). https://doi.org/10.1039/C7CS00205J

The Characterization Analysis of Graphene

Rupali Gupta, Dharmendra Kumar Yadav, Sasanka Deka, and Vellaichamy Ganesan

Abstract This chapter describes various state-of-the-art fabrication routes for high-quality graphene included: chemical vapour deposition (CVD), mechanical exfoliation, chemical exfoliation, electrochemical exfoliation, arc discharge, epitaxial growth, and pyrolysis. CVD is a widely used technique for growing high-quality graphene films on metal catalyst substrates, and copper foil has shown promising results. Mechanical exfoliation involves peeling graphite flakes from highly oriented pyrolytic carbon (HOPG) platelets using Scotch tape, resulting in single-layer graphene. Chemical exfoliation has two methods: solution-assisted and low-temperature chemical exfoliation. Electrochemical exfoliation involves the intercalation and exfoliation of graphite into graphene nanosheets through electrolyte solutions. Arc discharge is a plasma deposition technique for synthesizing high-quality graphene sheets using alternating current arc-discharge processes. Epitaxial growth involves growing single-layer or multilayer graphene on a SiC substrate using high-temperature sublimation growth. Pyrolysis is a 6-step process of poly(methyl methacrylate) composite that results in carbon derivatives that dissolve in the Ni catalyst surface, resulting in the epitaxial growth of graphene. Each method has its unique features, advantages, and disadvantages, making them suitable for different applications. For example, mechanical exfoliation remains one of the most reliable ways of producing high-quality graphene and has led to the discovery of graphene's extraordinary physical properties. Chemical exfoliation can produce graphene on a large scale, and electrochemical exfoliation is effective in creating biocompatible and fluorescent carbon nanomaterials for biological labelling and imaging. CVD as well as epitaxial growth can produce high-quality graphene films, and pyrolysis

R. Gupta
Department of Chemistry, M. M. Mahila College, Veer Kunwar Singh University, Ara, Bihar, India

D. K. Yadav (✉) · S. Deka
Department of Chemistry, University of Delhi, Delhi 110007, India
e-mail: dharmendra87chem@gmail.com

V. Ganesan
Department of Chemistry, Institute of Science, Banaras Hindu University, Varanasi, UP 221005, India

© The Author(s), under exclusive license to Springer Nature Singapore Pte Ltd. 2023
R. T. Subramaniam et al. (eds.), *Graphene*, Engineering Materials,
https://doi.org/10.1007/978-981-99-1206-3_6

produces graphene with a high degree of graphitization. The choice of the appropriate technique is crucial for specific applications.

Keywords Graphene · CVD · Exfoliation · Arc discharge · Pyrolysis

1 Spectroscopic and Microscopic Approaches

Graphene, a single atomic thick sheet with a hexagonally arranged layer of sp^2 hybridized C atom, has remarkably revolutionized the field of nanotechnology [1, 2]. It has provided researchers all over the world with myriads of opportunities to explore its various properties leading to its diverse range of applications. Various other derivatives of graphene, for example, graphene oxide (GO) and reduced graphene oxide (rGO) have been equally explored [1–5]. Graphene, along with its derivatives have inherited many unique properties such as excellent electronic conductivity, exceptional mechanical strength, high thermal stability, optoelectronic properties, etc. which offers their widespread employability in chemical sensors, biomedical systems, flexible electronic devices, touch screens, batteries, fuel cells, conductive films, high-frequency circuits, etc. The synthesis approach for graphene and its derivatives involves versatile methods such as chemical vapor deposition (CVD), chemical exfoliation of graphite, and thermal, photo, or electrochemical reduction of GO. In order to investigate the various amazing properties of graphene, it becomes very much essential to dig deep into its characteristic aspects [5]. Various standard protocols and a large number of characterization techniques are available making it quite a cumbersome task to compare different studies [1–5]. To simplify this comparison, the study of the characteristic features of graphene and its derivatives can be best categorized into the study of their composition and morphological features [2–4]. The compositional characterization deals with the atomic composition of the constituents i.e., atomic carbon/oxygen ratio whereas the morphological characterization provides information about the thickness of the sheet, specific surface area, bending elasticity, lateral dimensions, etc.

Furthermore, various techniques have been employed as characterization tools for depicting the compositional and morphological attributes of graphene and its derivatives [1–5]. Of these, the following characterization techniques are utilized for compositional study:

(i) Inductively coupled plasma mass spectrometry (ICP-MS);
(ii) X-ray photoelectron spectroscopy (XPS);
(iii) Elemental analysis; and
(iv) X-ray diffraction.

X-ray photoelectron spectroscopy is a spectroscopic technique used for chemical analysis. It is a popular tool employed by material chemists for analyzing surface chemistry by measuring elemental composition as well as the chemical and electronic state of the atoms present in a material [3]. This technique measures the number and kinetic energy of the emitted photoelectrons, generated by irradiating the sample surface with X-rays. The quantitative estimation of elements and the C/O ratio of GO is commonly determined using XPS. Moreover, XPS is of great utility when it comes to the characterization of heteroatom-doped graphene-based material. The most useful function of this technique for graphene and its derivatives is to confirm the occurrence of foreign atoms on graphene by XPS survey spectra.

In addition, XRD is another the most efficient and common technique employed for determining the crystallographic structure of a material [1, 2]. The working principle for XRD involves irradiating the material to be characterized with X-rays and then measuring the intensity of the X-rays which are scattered with respect to the scattering angle. This study is used to get an idea about the stacking of graphene and GO, quantitative estimation of the degree of exfoliation, etc. The prime role of XRD in characterizing graphene and its derivatives is to confirm the reduction of GO to graphene. For instance, the XRD spectra of pristine graphite show a diffraction peak at the 2θ value of 26.52° (002), corresponding to an interlayer distance of 0.34 nm as calculated using Bragg's equation. However, in GO, the diffraction peak at 26.52° disappears and a new peak appears at a lower 2θ value. Further, upon reduction of GO, the peak at lower the 2θ value disappears indicating the formation of reduced graphene oxide.

Moreover, the following are the well-established techniques that are used to study the morphological aspects:

(i) Raman spectroscopy;
(ii) Scanning electron microscopy (SEM);
(iii) Optical microscopy;
(iv) Scanning probe and tunneling microscopy;
(v) Transmission electron microscopy (TEM) and high-resolution TEM (HR-TEM); and
(vi) Atomic force microscopy (AFM).

In this chapter, we provide an overview of several characterization strategies employed for studying the morphological aspects of graphene and its derivatives, regarding their mode of operation, and also a brief literature review of the various techniques employed for the successful characterization of graphene.

We start with the Raman spectroscopic studies for graphene and its derivatives, which gives information about the number of layers and defects or disorders in the crystal structure. Next, SEM is discussed, which is used for studying the surface morphology of graphene. Next, the optical microscopic studies are described. Here, a detailed analysis of literature related to scanning probe and tunneling microscopy studies on graphene and its derivatives are provided. Scanning probe and tunneling microscopy provides a picture of morphology and electronic structure. A descriptive study on TEM and HR-TEM which are majorly employed in quantitative estimation of graphene layers by getting cross-section images is also discussed. Subsequently, AFM is discussed, whereby this technique also can be used to estimate the number of graphene layers on the basis of the interlayer distance.

1.1 Raman Spectroscopy

Raman spectroscopy is a vibrational spectroscopic technique that detects even minor changes in geometric structure and bonding within the molecule. Raman spectroscopy is a reflecting, noncontact technique and does not require any tedious sample preparation procedures [6]. High-quality Raman spectra can be measured quickly. This property is beneficial for studying distinct allotropes of carbon, where the differences between them are the relative positions of their carbon atoms and the nature of their bonding. As a result, Raman spectroscopy has become an essential technique in laboratories for researchers who are working on carbon nanomaterials research.

Analysis, monitoring, and refining of graphene synthesis process have got an important role in research and industry applications [7]. Raman spectroscopy has emerged as a significant characterization tool in graphene research, as it allows researchers to get critical material properties locally and non-invasively [8]. The principles of graphene's Raman spectrum as well as how to use Raman spectroscopy for graphene characterization are covered in this chapter. In this chapter, we will show how the Raman spectrum can be used to extract important properties of graphene samples, such as doping, strain, defect density, and layer number.

At the first look, the Raman spectrum of graphene or graphite appears to be rather straightforward, with only two main bands and a few minor bands. These bands reveal a lot about the material.

The G-band which appears around 1582 cm^{-1} (graphite) and the G'-band which appears around 2700 cm^{-1} with laser excitation at 2.41 eV are the most conspicuous characteristic in the Raman spectra of monolayer graphene [9]. The G-band depicts that the planar sp^2 hybridized bonded carbon is responsible to fabricate graphene. However, D-band represents the disorder-induced band which can be seen in a disordered sample or near the edge of a graphene sample. Moreover, the D-band is observed for all sp^2 hybridized disordered carbon materials at a frequency of about half of the G'-band (approximately 1350 cm^{-1} with 2.41 eV laser excitation) [10]. Also known as the disorder or defect band, the D-band depicts a ring breathing mode from sp^2 carbon rings, although in order to be active, it should be near a graphene edge or a defect [6]. A single phonon lattice vibrational action produces the band [10, 11]. The band is typically very weak in graphite as well as in graphene. The presence of a significant D-band infers that the material contains a large number of flaws. The intensity of the D-band corresponds to the number of faults in the material [12]. In graphene, the G-band is produced from a regular first-order Raman scattering process. Some publications prefer to refer to the G'-band as the 2D-band because its frequency is roughly twice that of the D-band [10–12]. The second-order Raman spectra of graphene show a two-phonon band without any disorder or flaws [10–12]. This band is also employed to detect the thickness of the graphene layer; however, unlike the G-band, the differences between single and bilayer graphene in this band are more complicated than a simple band shift. Furthermore, the Raman spectra of single-layer graphene (Fig. 1) and multilayer graphene (Fig. 2) are discussed below.

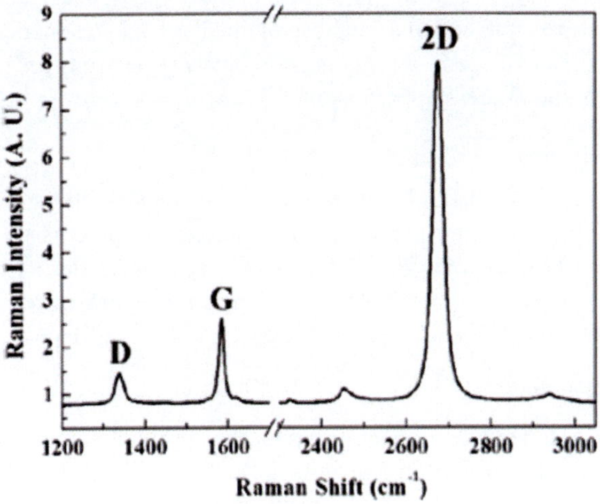

Fig. 1 Raman spectrum of monolayer graphene measured with a 532 nm laser excitation. (Reproduced with permission from Heo et al. [16])

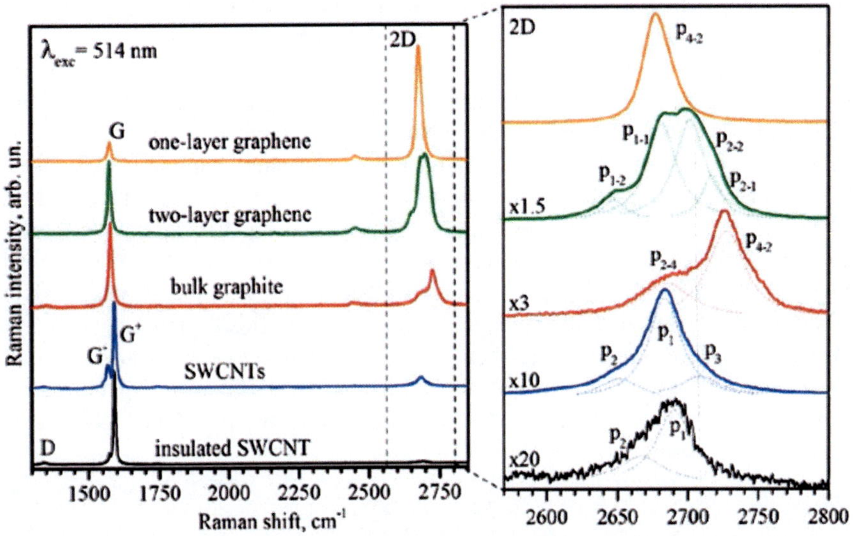

Fig. 2 Raman spectra of mono, bi, and bulk layers of graphene. (Reproduced with permission from Stubrov et al. [17])

(i) Monolayer graphene

The G- and G'-bands are shown in Fig. 1 for the monolayer graphene sample recorded with a laser excitation of 2.41 eV [13–16]. At room temperature, the G'-band for single-layer graphene has a single Lorentzian feature showing a full width at half maximum (FWHM) of 24 cm^{-1} [14]. The relative intensity of the G'- and G-bands is striking in this spectrum and this can be explained as triple resonance process [15]. Due to the linear E(k) dispersion, this is an exclusive spectrum for sp^2 carbons, in which all steps in the process are resonant [13–15].

(ii) Bilayer graphene

The electrical and phonon bands break into two components with specific symmetries in bilayer graphene resulting in four separate scattering patterns [16–18]. The G' spectrum has four Raman peaks due to these four separate scattering processes. The Raman spectra of bilayer graphene samples having an AB stacking pattern were fitted with four Lorentzians [17], each with FWHM of 24 cm^{-1}, as depicted in Fig. 2.

(iii) Tri-layer graphene

It can be demonstrated on the basis of group theory approach for a tri-layer graphene that the number of allowable Raman peaks in the G'-band is fifteen [19]. However, experimentally, it is found that the G' spectrum has fitted with a minimum of six peaks with FWHM of 24 cm^{-1}. On an important note, the G'-band which results from the monolayer graphene always has the same FWHM of 24 cm^{-1}. As more graphene layers are added, the symmetry of the G'-band decreases, causing it to break into numerous overlapping modes. For the layer thicknesses of less than four layers, these unique band shape changes allow the successful distinction between single and multilayer graphene [20]. However, the ratio of the peak intensity of the G'-and G-bands can also be used to identify single-layer graphene. For high-quality single-layer graphene, the ratio of these bands, $I_{G'}/I_G$ is greater than or equal to 2 [21]. This ratio is frequently used to validate that a graphene sample is defect-free [21]. Raman spectra can be used to find the disorder level in graphene based on the peak intensities ratio i.e., I_D/I_G. I_D/I_G exhibits two distinct behaviors as graphene disorder rises [21, 22]. This ratio increases in a "low" defect density regime, because a larger defect density causes more elastic scattering. This persists until a "high" defect density regime is reached, after which the I_D/I_G ratio begins to drop as the defect density increases resulting in a more amorphous carbon structure which attenuates all Raman peaks. The terms "nanocrystalline graphite" and "primarily sp^2 amorphous carbon" are used to describe these two phases. However, one can calculate the degree of edge defect based on Eq. 1 [23, 24]:

$$\frac{I_D}{I_G} = \frac{C'(\lambda)}{L_D} \tag{1}$$

where, λ is the Raman excitation wavelength, $C'(\lambda) = (2.4 \times 10^{-10}\ \mathrm{nm}^{-3})\ \lambda^4$.

The strain on the graphene which is commonly generated by the mismatch of the lattice with the underlying substrate can also cause the contraction and expansion of the lattice [25]. As with doping effects, extending the lattice reduces phonon energies causing a red shift of the Raman spectra [25]. The intensity of the G' peak is similarly decreased when graphene is doped with other elements/materials [26]. If the strain is uniaxial as in bending, the G peak will split into two distinct features, which correspond to the splitting of the vibrational mode into one parallel to the curve and one perpendicular to it. The concentration of electron and lattice strain in graphene sample can be described using a variety of methods. Temperature also plays a role to tune the Raman spectra. The peak positions of the Raman spectra shift in response to changes in temperature [27]. It is due to the enhanced anharmonic coupling of phonons and greater thermal expansion in the lattice. A rise in temperature causes a linear red shift in G'- and G-bands. In this way, Raman spectroscopy can be utilized to determine the graphene structure.

1.2 Scanning Electron Microscopy (SEM)

An excellent tool developed to study the surface morphology of graphene and its derivatives is SEM. It is based on the principle of producing the image of the sample upon scanning it with a beam of high-energy density electrons. Upon their interaction with the sample, secondary electrons, backscattered electrons, and X-rays are produced. Furthermore, SEM technique easily manages to get information about the topology as well as the lateral dimension but fails to quantify the thickness of graphene sheet(s). For instance, Xu et al. have shown that the SEM image of GO having 2D nanosheets like morphology with a lateral thickness of about 20 nm (Fig. 3a and b) [28] whereas, in another study, Ren et al. have reported the SEM images of GO which were observed to have a rough texture, irregular edges with crumples (Fig. 3c) [29]. In addition, Tarcan and his co-workers have depicted the SEM images of stacked thin films of reduced GO having designed patterns developed by laser writing method and porous rGO developed by laser irradiation (Fig. 3d–f) [30]. The cross-sectional image of the stacked sheets of GO can be seen comprising thin films stacked one upon another whereas the porous rGO exhibited pores like structure.

Alternatively, Bai et al. have also reported the SEM images of a composite of GO incorporated with hydrogenated carboxylated nitrile-butadiene rubber (GO/HXNBR) [31]. The cross-sectional image of GO/HXNBR has been reported at lower and higher magnifications visualizing a clearer image of the composite at higher magnifications [31]. Moreover, Khan et al. have very well demonstrated the morphological difference between the GO formed by chemical reduction method and thermal exfoliation technique by employing SEM technique. It can be well established by the SEM study that rGO with a higher aspect ratio can be produced via thermal exfoliation as compared to the chemical reduction method (Fig. 4a and b) [32].

Choi et al. has described the synthetic approach to get the porous graphene aerogels via hydrothermal and electrochemical reduction of GO [33]. A comparative study of the surface morphology of GO synthesized by hydrothermal and electrochemical reduction routes was done using field emission scanning electron microscopy (FESEM) which depicted a clear difference between the two. The FESEM image of graphene aerogels (hydrothermal reduction of GO) showed a highly porous structure having an interconnected 3D graphene network. In another report, a detailed

Fig. 3 SEM images of GO sheets: **a** Low and **b** high magnification. (Reproduced with permission from Xu et al. [28]); **c** SEM image of GO (Reproduced with permission from Ren et al. [29]); Tilted **d** and cross-sectional **e** and **f** images of thin films of stacked GO and porous rGO (Reproduced with permission from Tarcan et al. [30])

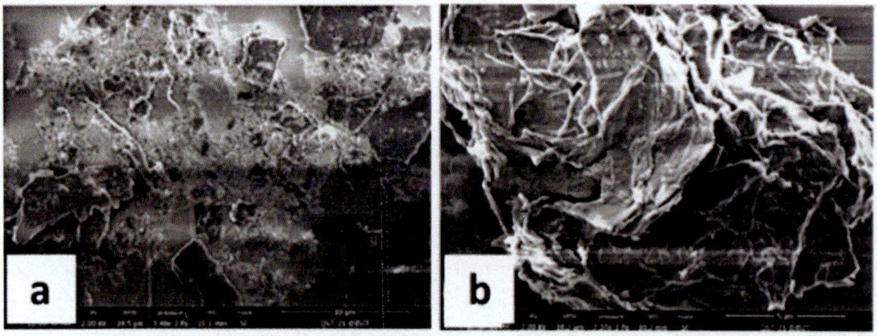

Fig. 4 FE-SEM images of **a** chemically reduced and **b** thermally exfoliated GO. (Reproduced with permission from Khan et al. [32])

synthesis of rGO and its composite with nickel phthalocyanine (rGO-NiPc) was reported [34]. Here, the SEM images of the two illustrated a considerable difference with rGO having a uniform laminar structure and that of rGO-NiPc displayed a denser morphology [34]. Moreover, Wang et al. demonstrated a very intriguing implication of the SEM technique [35]. Here, the growth of graphene on copper by chemical vapor deposition in a modified environmental SEM chamber was investigated. This study provided an overall clarity of the CVD process covering each step, for instance, substrate annealing, graphene growth, and cooling respectively.

In addition, a detailed study of the external factors (temperature, pressure, and surrounding) with respect to the dynamics of the substrate and their influence on the growth of graphene and the nucleation process has been previously conducted. Illustrated in Fig. 5 are the SEM images captured at different times during the nucleation process [35]. With the passage of time, the appearance of graphene sheets and their growth are visible. The exact clear image of the nucleation process could not be obtained, however, the appearance of new flakes was observed at a size of 3 nm.

The synthesis of rGO and oxovanadium salen embedded rGO (rGO-VO (salen)) and the morphology of the two by SEM were also reported [36]. The SEM image of rGO clearly depicts randomly associated thin layers whereas the rGO-VO (salen) exhibited densely coagulated spots which could be attributed to the encapsulation of VO (salen) on the rGO (Fig. 6a and b) [36].

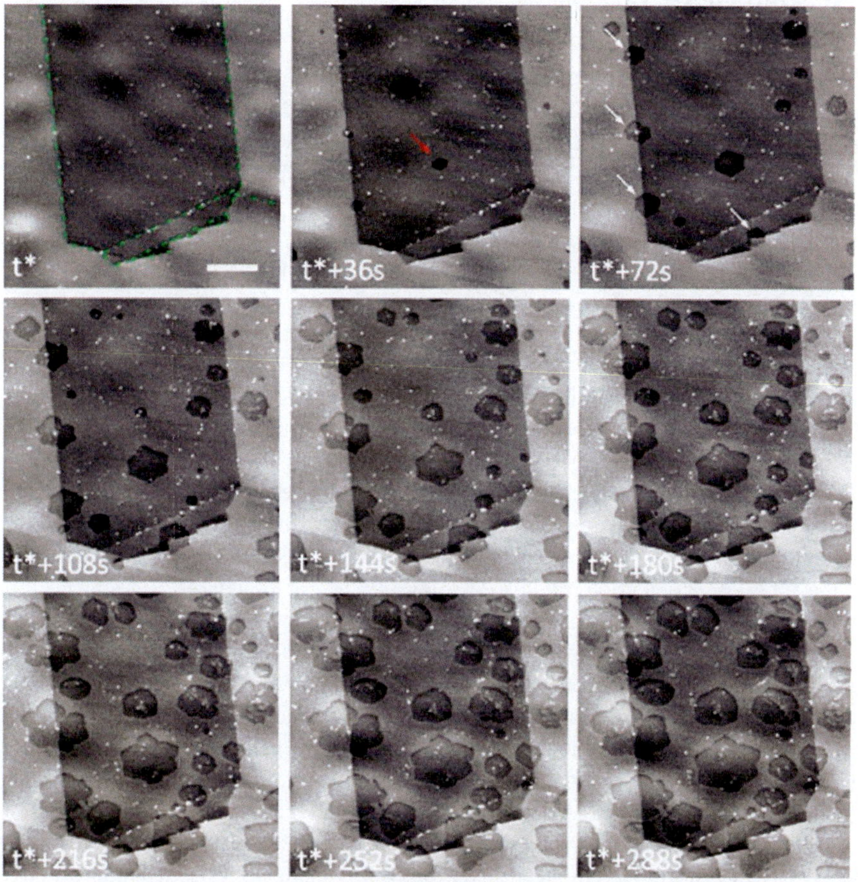

Fig. 5 SEM images taken at 1000° C. White color arrows correspond to the nucleation process taking place at the grain boundaries. (Reproduced with permission from Wang et al. [35])

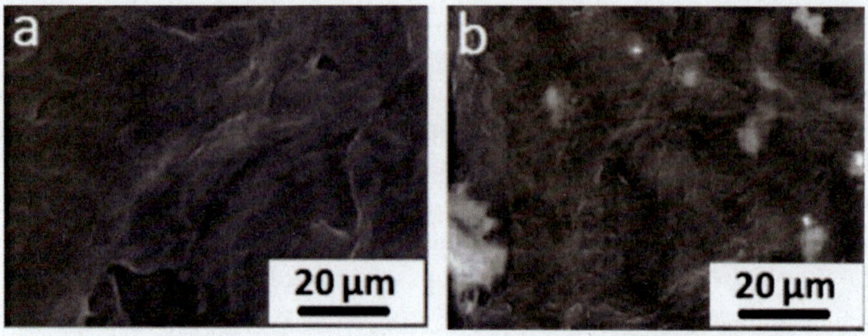

Fig. 6 SEM images of rGO (**a**) and oxovanadium salen integrated rGO (**b**). (Reproduced with permission from Sonkar et al. [36])

1.3 Optical Microscopy

Optical microscopy is also a useful technique employed for graphene characterization. Though graphene is only a single atom layer thick, it can be viewed using an optical microscope. The first finding of graphene was achieved using a standard optical microscope [37]. The procedure is straightforward in which exfoliated graphene obtained by mechanical peeling is deposited on Si/SiO_2 and viewed directly under the optical microscope [38, 39].

The color and contrast of graphene layers depend on the number of graphene layers [39]. When graphene is deposited on a clean silicon substrate it cannot be seen with an optical microscope. For obtaining the optical image of graphene, oxide layer thickness plays an extremely critical role [38]. Since graphene has several optical path diffraction and interference effects which depend upon the oxide layer thickness on the silicon substrate, graphene has a distinct color and contrast [37–39]. Graphene can also be seen using optical microscopy on other substrates such as Si_3N_4 or polymethyl methacrylate with the right thickness [40]. Color and contrast can be matched with the number of graphene layers using AFM [38]. Furthermore, the number of layers of graphene can be calculated based on the color and contrast in the subsequent detection. Figure 7 displays microscopic pictures of graphene and thin graphite sheets with various layers [39]. Among these, the variety of monolayer graphene differs little from that of the silicon substrate, indicating the excellent transparency of graphene [39]. A thermal annealing procedure can be also used to optically characterize graphene domains directly via CVD grown graphene on metallic substrates such as nickel, copper, etc. (Fig. 8) [41]. For instance, thermal annealing converts the naked copper without graphene to copper oxides, increasing the interference color contrast between oxides and pure copper and making graphene visible under an optical microscope [42]. Moreover, heat treatments can also be used to identify the multilayer graphene stacks on copper using direct optical microscopy.

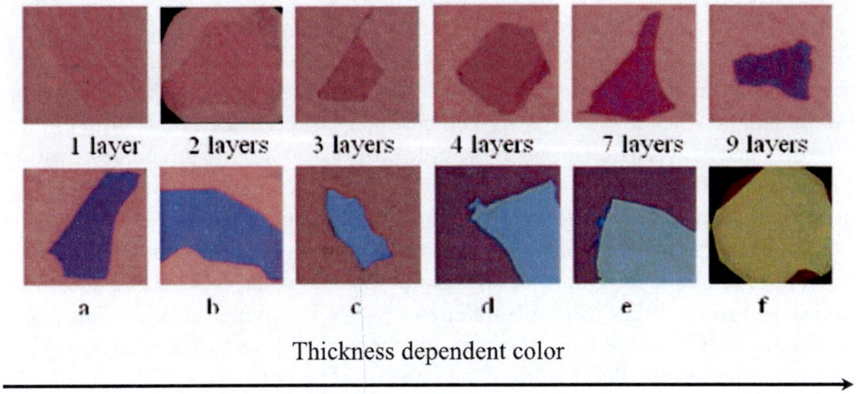

Fig. 7 Optical microscopic analysis of the various thickness of graphene. (Reproduced with permission from Ni et al. [39])

Fig. 8 Optical microscopic image of graphene domains directly grown on a copper substrate. (Reproduced with permission from Jia et al. [41])

1.4 Scanning Probe and Tunneling Microscopy

In this section, the basics of graphene characteristics through scanning probe and tunneling microscopy are discussed. Scanning tunneling spectroscopy (STM) is used to characterize the single layer of graphene which is deposited or grown on a variety of substrates as well as the effects of different substrates on graphene characteristics. Since it directly examines the electrical characteristics at the surface, scanning probe microscopy (SPM) is an appropriate tool for studying low-dimensional materials. Numerous techniques have been used for the synthesis of graphene [43–45]. Various properties of the surface can be examined on the basis of interaction between the probe and the surface. SPM has made significant contributions to our understanding of graphene. The essential operational principles of primary types of SPM that have been used for the study of graphene will be further discussed. After the invention of the STM, Gerd et al. reported the atomic-level characterization of monolayers of carbon on Ru (0001) and Ru (1120) [46], Ni (111) [47], and Pt (111) [48] respectively [49].

In addition, a STM is equipped with a sharp metallic tip and a conducting sample separated by a few angstroms, which aids in capturing the image of atoms on the surface. The tunneling of electrons causes an electric current to flow between the tip and the sample when a voltage is applied between them. The current is proportional to the distance between the tip and the sample [50]. The atomic-scale characterization of the surface of the material can be achieved by raster scanning. Scanning tunneling spectroscopy can also provide information about the sample's densities [51].

STM analyses on a range of graphene samples have revealed the honeycomb lattice structure of single-layer graphene on various substrates like SiO_2, metal substrate, graphite substrate, h-BN, SiC substrate, etc. [52–56]. Among them, few measurements have been described. For instance, graphene can be investigated on a graphite substrate, after electrically detaching the uppermost layer from the bulk graphite. Li et al. was the first researcher to look into this type of graphene sample [54]. Moreover, a highly oriented pyrolitic graphite (HOPG) is a type of polycrystalline graphite that was formed from pyrolytic graphite. In HOPG, the surface was decoupled from the bulk, but this was not the case in Kish graphite [57]. (It is a single crystal produced from the carbon-saturated iron melt.) However, STM pictures of decoupled graphene flakes were reported in another investigation on similar graphene samples, showing the presence of hexagonal lattice on the topmost layer and triangular lattice on the layer below the topmost layer. A lengthy ridge on the graphite surface divided the layers [58].

Furthermore, the hexagonal structure of the exfoliated monolayer graphene on the SiO_2 substrate was observed to be visible through STM pictures [53]. Here, the images showed the corrugations of the underlying SiO_2 substrate, partially confirming the presence of graphene on SiO_2 [59]. Figure 9 illustrated a large-scan-area STM picture of graphene on the copper surface which exhibited the hexagonal structure of single-layer graphene [58–60]. In other words, the hexagonal lattice confirmed the electronic input from both sub-lattices of carbon atoms, in contrast to simply one sub-lattice contributing to the image, as observed in bilayer graphene or 3D graphite.

1.5 Transmission Electron Microscopy (TEM)

Interestingly, TEM has proved to be an excellent and powerful tool in the field of graphene research and development. In TEM, a high-energy beam of electrons is used which interacts with the atoms of the sample and the resulting data can be used to derive various information related to the thickness of the layer, crystal structure, dislocations, grain boundaries, etc. As the electrons have a very small wavelength, the resolution obtained in the case of TEM is very high which aids in analyzing the shape and size of the nanomaterials efficiently. TEM is also used to investigate the growth of layers and defects in a crystal structure. TEM analysis has been very proficiently employed for the study of graphene and graphene-based nanocomposites. For instance, Xu et al. have reported the synthesis of a nanocomposite composed

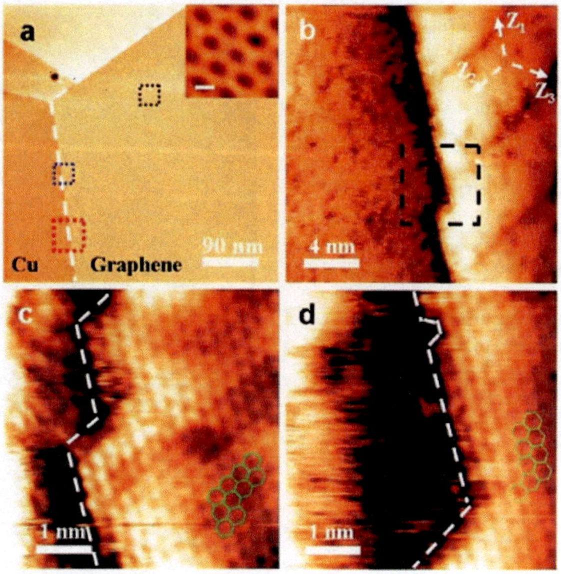

Fig. 9 Surface topography of graphene layers on the copper surface displaying two zigzag-oriented edges (dashed white lines). (Reproduced with permission from Tian et al. [58])

of an array of nanowires made of polyaniline on GO sheets and employed them for supercapacitor application [28]. Here, the lower and higher magnifications of TEM images of GO showed layers having curled appearance (Figs. 10a and b) [28]. Alternatively, Bao et al. synthesized graphene by employing pressurized oxidation and multiplex reduction and its nanocomposite with poly(lactic acid) using the master batch-based melt blending method [61]. The TEM images of GO and graphene (Figs. 10c and d) further confirmed that the layer size of GO and graphene was much smaller compared to those materials obtained by ultra-sonication method [61]. Also, Fig. 10(e–g) illustrated the TEM representations of graphene-poly(lactic acid) nanocomposite [61]. Exfoliation of the graphene nano-layers indicated a molecular degree of dispersion.

TEM characterization of the materials has also been reported for the rGO and oxovanadium salen embedded rGO (rGO-VO (salen)) [36]. The TEM image of rGO illustrated a 2D crumpled sheet-like appearance with few shrinkages (Fig. 11a) [62]. The presence of VO (salen) on rGO was confirmed by the TEM studies which further revealed the nanocrystals, consisting of a size of approximately 200 nm (Fig. 11b) [62]. In addition, Li et al. investigated on the non-covalent surface modification of pristine graphene using a single step reaction between sodium carboxymethyl cellulose (CMC) and pristine graphene [62]. Here, the TEM studies revealed that post-modification with CMC the pristine graphene sheets displayed a better dispersion in water. TEM images of pristine graphene sheets displayed tightly agglomerated large sheets indicating its poor dispersion, but after modification with CMC, the TEM image displayed well-dispersed small nanosheets in water (Fig. 11c and d) [62].

Alternatively, in another study, Jana et al. observed the TEM images which further revealed various details about graphene-based nanocomposite [63]. Here,

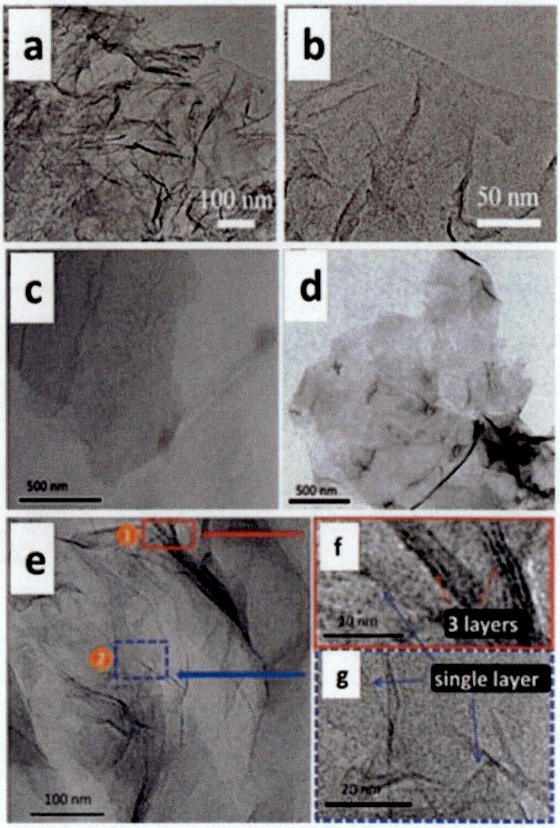

Fig. 10 Low (**a**) and high magnification (**b**) TEM images. (Reproduced with permission from Xu et al. [28]), TEM images of GO (**c**) and graphene (**d**) (Reproduced with permission from Bao et al. [61]), and TEM image of graphene in poly(lactic acid) (**e**). Images (**f**) and (**g**) are the high-resolution images of the area marked as '1' and '2' respectively, in the image (**e**) (Reproduced with permission from Bao et al. [61])

a solvothermal approach for synthesizing europium-incorporated ZnO-chemically converted graphene nanocomposites (ZEG) was reported. It was concluded from the TEM studies (Fig. 12) that with the increased concentration of Eu, a reduction in the agglomeration of the composite nanoparticles was observed [63]. The composite nanoparticles were found to be monodispersed and quasi-spherical in shape. As the Eu level increased, the trend in the change in average particle size as determined by histograms created from different TEM images was found to be similar to the crystallite size of h-ZnO as determined by the XRD study.

In addition, Dresselhaus et al. reported on the various edge and ribbon-like structures of graphene produced experimentally or theoretically [64]. Nanoribbons made up of edged graphene can be made in a variety of methods [64–66]. Despite the fact that lithographically created graphene nanoribbons are routinely employed in research labs, chemically formed graphene nanoribbons have also been studied. As the lithographically or chemically produced graphene nanoribbons are narrower and produced in low quantities, the CVD technique was employed as an alternative to producing a large number of graphene nanoribbons [19]. Figure 13a illustrated the lower magnification TEM image of suspended graphene nanoribbons produced by

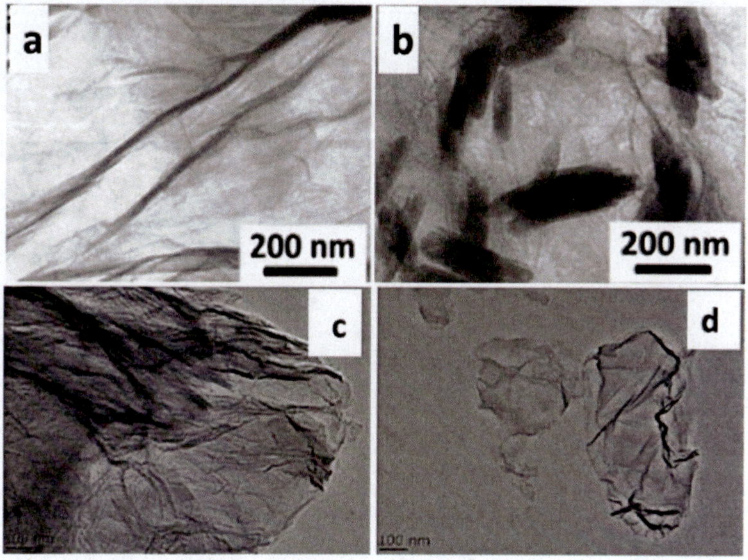

Fig. 11 TEM pictures of rGO (**a**) and oxovanadium salen embedded rGO (**b**). (Reproduced with permission from Sonkar et al. [36]). TEM pictures of pristine graphene sheets (**c**) and pristine graphene sheets modified with CMC (**d**) (Reproduced with permission from Li et al. [62])

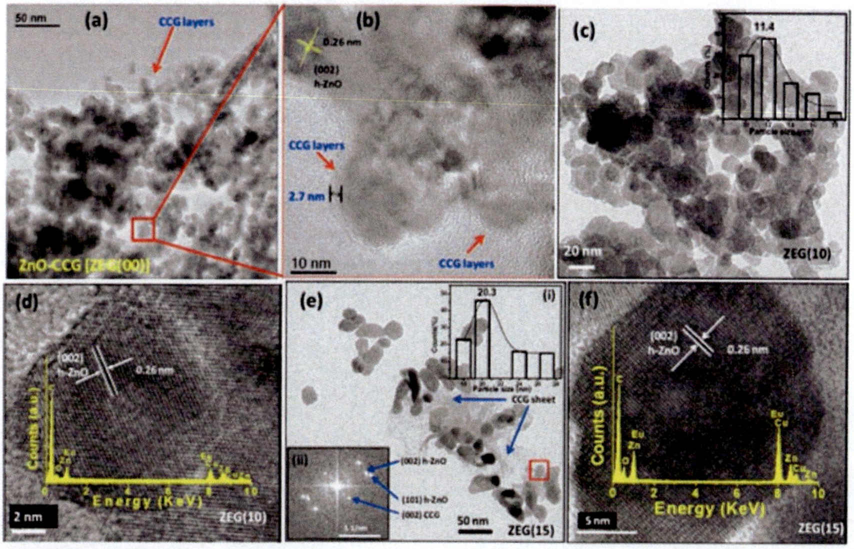

Fig. 12 TEM images of Eu incorporated ZnO-chemically converted graphene nanocomposites (ZEG) (Having different concentrations of Eu). (Reproduced with permission from Bera et al. [63])

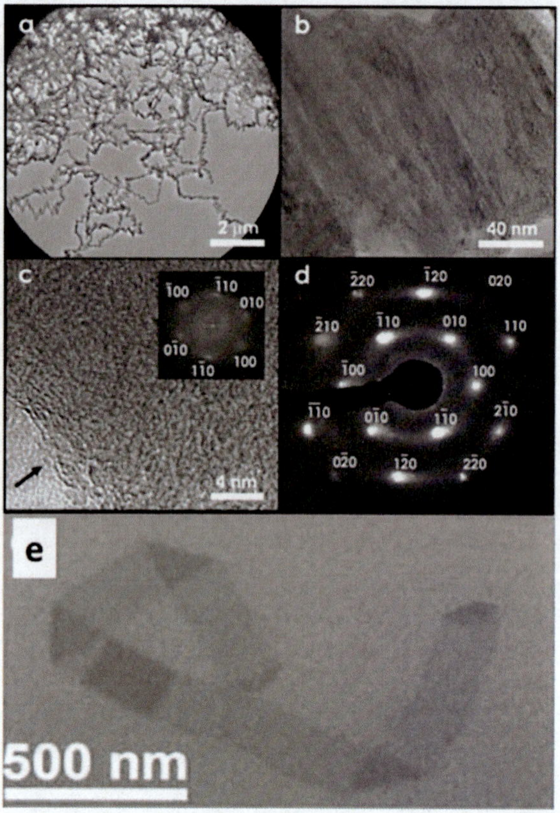

Fig. 13 Low (**a**) and high (**b**) magnification TEM images of graphene nanoribbons, HR-TEM image of a ribbon edge (**c**), the indexed electron diffraction pattern of an individual thin graphene nanoribbon (**d**). (Reproduced with permission from Delgado et al. [65]), and TEM image (**e**) of a graphene nanoribbon displaying bends (Reproduced with permission from Jia et al. [64])

the CVD method [64, 65]. Furthermore, another method employed to obtain graphene nanoribbon is unzipping the carbon nanotubes along their longitudinal direction [64]. Illustrated in Fig. 13b, Higginbotham et al. revealed the TEM image of graphene nanoribbons having various bends produced by unzipping nanotube [64].

1.6 Atomic Force Microscopy

Scanning probe microscopy was utilized for the first time to image the exfoliated graphene flakes by Novoselov et al. [60]. The most frequent type of scanning probe microscopy technique used to characterize graphene is AFM. AFM has played an important role to probe and image the thickness of 2D materials [61]. AFM creates images by scanning with a cantilever across a sample's surface. The cantilever's sharp nanoscale tip makes contact with the surface, bending it and changing the amount of laser light reflected into the photodiode [60, 67, 68]. The cantilever height is then modified to restore the response signal resulting from tracing the surface with the

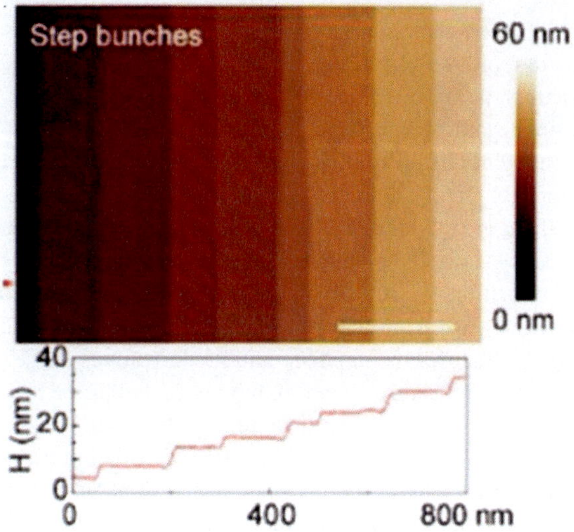

Fig. 14 AFM image of graphene grown on Cu foil and its height profile on red point. (Reproduced with permission from Deng et al. [71])

observed cantilever height. In 1999, Ruoff's group demonstrated one such method for graphite by plasma etching tiny pillars shaped into highly oriented pyrolytic graphite with the AFM tip [69]. This method is sensitive and time-taking. The step height for each graphene layer varied from 0.34 to 1.0 nm, which was well within the contemporary AFM detection limits. It was due to the differences in interactions as a result of which the substrate graphene and AFM probe-graphene are thought to be related to variations in substrate surface energy, graphene structure, and sample preparation [70]. Graphene grown on Cu (111) foil is characterized using AFM, as depicted in Fig. 14 [71]. AFM is also used to test the mechanical properties such as elastic stiffness and fracture strength of ultra-flat graphene [70].

2 Conclusion

With unique chemical composition and structure and extraordinary electronic, optical, thermal, and mechanical properties, graphene and its derivatives provide a range of utility. A large number of fabrication methods have been developed to obtain the desired form of graphene and its derivatives. In order to examine and study its wonderful properties, various characterization techniques have also been established. Proposing a common characterization strategy for graphene is difficult due to its complicated nature. In some applications, some properties of graphene are more significant than others. Several basic graphene characteristics, on the other hand, appear to be relevant in practically all research. The degree of disorder of the graphene sheets, graphene stacking, the lateral dimensions of the graphene, and the functional groups contained in/on the graphene are all examples. As a large number of characterization techniques are available for examining graphene and its derivatives, it becomes equally essential to categorize these techniques. These characterization techniques are classified into two major groups based on compositional and morphological characterization. Whereas compositional characterization techniques provide quantitative information regarding the atomic constituents, morphological techniques enlighten the lateral dimension, thickness, and specific surface area. Inductively coupled plasma mass spectrometry, X-ray photoelectron spectroscopy, and elemental analysis are a few techniques employed for studying the compositional aspects of graphene and its derivatives. In this chapter, we provide a summary of various characterization techniques employed for studying the morphological features of graphene and its derivatives namely Raman spectroscopy, SEM, optical microscopy, scanning tunneling and probe microscopy, TEM, and AFM. An exhaustive study of literature reports explains the fact that not all characterization techniques are utilized uniformly. Many techniques are utilized more commonly than others. These approaches are interrelated when used together, assessing sheet spacing, chemical functionality, height, and lateral dimensions of graphene and its derivatives.

Acknowledgements Dharmendra Kumar Yadav thanks the UGC for Dr. D.S. Kothari Postdoctoral Fellowship (F.4-2/2006 (BSR)/CH/20-21/0244), India. VG thanks the IoE grant of Banaras Hindu University (Scheme number: 6031) for funding.

References

1. Huang, Y.Y., Wu, J.: Preparation and characterization of graphene oxide/polyaniline/carbonyl iron nanocomposites. Materials **15**, 484–500 (2022). https://doi.org/10.3390/ma15020484
2. Tambe, P.: Synthesis and characterization of acid treated reduced graphene oxide. Mater. Today Proc. **49**, 1294–1297 (2022). https://doi.org/10.1016/j.matpr.2021.06.381
3. Ding, Z., Tang, Y., Zhu, P.: Reduced graphene oxide/cellulose nanocrystal composite films with high specific capacitance and tensile strength. Int. J. Biol. Macromol. **200**, 574–582 (2022). https://doi.org/10.1016/j.ijbiomac.2022.01.130
4. Kant, R., Sharma, T., Bhardwaj, S., Kumar, K.: Structural, electrical and optical properties of MgO-reduced graphene oxide nanocomposite for optoelectronic applications. Curr. Appl. Phys. **36**, 76–82 (2022). https://doi.org/10.1016/j.cap.2022.01.008
5. Allen, M.J., Tung, V.C., Kaner, R.B.: Honeycomb carbon: a review of graphene. Chem. Rev. **110**, 132–145 (2010). https://doi.org/10.1021/cr900070d
6. Kakihana, M., Osada, M.: Raman spectroscopy as a characterization tool for carbon materials. Carbon Alloy. 285–298 (2003). https://doi.org/10.1016/B978-008044163-4/50018-8
7. Wilhelm, H., Lelausian, M., Rae, M.E., Humbert, B.: Raman spectroscopic studies on well-defined carbonaceous materials of strong two-dimensional character. J. Appl. Phys. **84**, 6552–6560 (1998). https://doi.org/10.1063/1.369027
8. Cancado, L.G., Pimenta, M.A., Neves, R.A., Ribeiro, M.G., Enoki, T., Kobayashi, Y., Takai, K., Fukui, K., Dresselhaus, M.S., Saito, R., Jorio, A.: Anisotropy of the Raman spectra of nanographite ribbons. Phys. Rev. Lett. **93**, 047403–047407 (2004). https://doi.org/10.1103/PhysRevLett.93.047403

9. Roscher, S., Hoffmann, R., Ambacher, O.: Determination of the graphene-graphite ratio of graphene powder by Raman 2D band symmetry analysis Anal. Anal. Methods **11**, 1224–1228 (2019). https://doi.org/10.1039/c8ay02619j
10. Pocsik, I., Hundhausen, M., Koos, M., Ley, L.: Origin of the D peak in the Raman spectrum of microcrystalline graphite. J. Non-Cryst. Solids **227**, 1083–1086 (1998). https://doi.org/10.1016/S0022-3093(98)00349-4
11. Ramsteiner, M., Wagner, J.: Resonant Raman scattering of hydrogenated amorphous carbon: evidence for π-bonded carbon clusters. Appl. Phys. Lett. **51**, 1355–1359 (1987). https://doi.org/10.1063/1.98677
12. Matthews, M.J., Pimenta, M.A., Dresselhaus, G., Dresselhaus, M.S., Endo, M.: Origin of dispersive effects of the Raman D band in carbon materials. Phys. Rev. B **59**, R6585–R6589 (1999). https://doi.org/10.1103/PhysRevB.59.R6585
13. Mafra, D.L., Samsonidze, G., Malard, L.M., Elias, D.C., Brant, J.C., Plentz, F., Alves, E.S., Pimenta, M.A.: Determination of LA and TO phonon dispersion relations of graphene near the dirac point by double resonance Raman scattering. Phys. Rev. B **76**, 233407–233411 (2007). https://doi.org/10.1103/PhysRevB.76.233407
14. Narula, R., Reich, S.: Double resonant Raman spectra in graphene and graphite: a two-dimensional explanation of the Raman amplitude. Phys. Rev. B **78**, 165422–165428 (2008). https://doi.org/10.1103/PhysRevB.78.165422
15. Ferrari, A.C., Robertson, J.: Raman spectroscopy of amorphous, nanostructured, diamond-like carbon, and nanodiamond. Philos. Trans. R. Soc. Lond. A **362**, 2477–2512 (2004). https://doi.org/10.1098/rsta.2004.1452
16. Heo, G., Kim, Y.S., Chun, S.-H., Seong, M.J.: Polarized Raman spectroscopy with differing angles of laser incidence on single-layer graphene. Nanoscale Res. Lett. **10**, 45–50 (2015). https://doi.org/10.1186/s11671-015-0743-4
17. Stubrov, Y., Nikolenko, A., Gubanov, V., Strelchuk, V.: Manifestation of structure of electron bands in double-resonant Raman spectra of single-walled carbon nanotubes. Nanoscale Res. Lett. **11**, 2–7 (2016). https://doi.org/10.1186/s11671-015-1213-8
18. Gruneis, A., Saito, R., Samsonidze, G.G., Kimura, T., Pimenta, M.A., Jorio, A., Filho, S.A.G., Dresselhaus, G., Dresselhaus, M.S.: Inhomogeneous optical absorption around the K point in graphite and carbon nanotubes. Phys. Rev. B **67**, 165402–165409 (2003). https://doi.org/10.1103/PhysRevB.67.165402
19. Malard, L.M., Guimaraes, M.H.D., Mafra, D.L., Mazzoni, M.S.C., Jorio, A.: Group-theory analysis of electrons and phonons in N-layer graphene systems. Phys. Rev. B **79**, 125426–125434 (2009). https://doi.org/10.1103/PhysRevB.79.125426
20. Lespade, P., Marchand, A.: Superlattice ordering in graphite-ICl single crystals and fibers. Carbon **22**, 325–333 (1984). https://doi.org/10.1016/0008-6223(84)90002-2
21. Ferrari, A.C., Meyer, J.C., Scardaci, V., Casiraghi, C., Lazzeri, M., Mauri, F., Piscanec, S., Jiang, D., Novoselov, K.S., Roth, S., Geim, A.K.: Raman spectrum of graphene and graphene layers. Phys. Rev. Lett. **97**, 187401–187405 (2006). https://doi.org/10.1103/PhysRevLett.97.187401
22. Lambie, S., Steenbergen, K.G., Gaston, N., Paulus, B.: Clustering of metal dopants in defect sites of graphene-based materials. Phys. Chem. Chem. Phys. **24**, 98–112 (2022). https://doi.org/10.1039/D1CP05008G
23. Tuinstra, F., Koenig, J.L.: Raman spectrum of graphite. J. Chem. Phys. **53**, 1126–1132 (1970). https://doi.org/10.1063/1.1674108
24. Lucchese, M.M., Stavale, F., Ferreira, E.H., Vilani, C., Moutinho, M.V., Capaz, R.B., Achete, C.A., Jorio, A.: Quantifying ion-induced defects and Raman relaxation length in graphene. Carbon **48**, 1592–1597 (2010). https://doi.org/10.1016/j.carbon.2009.12.057
25. Ni, Z.H., Yu, T., Lu, Y.H., Wang, Y.Y., Feng, Y.P., Shen, Z.X.: Uniaxial strain on graphene: Raman spectroscopy study and band-gap opening. ACS Nano **2**, 2301–2305 (2008). https://doi.org/10.1021/nn800459e
26. Pisana, S., Lazzeri, M., Casiraghi, C., Novoselov, K.S., Geim, A.K., Ferrari, A.C., Mauri, F.: Breakdown of the adiabatic Born-Oppenheimer approximation in graphene. Nat. Mater. **6**, 198–201 (2007). https://doi.org/10.1038/nmat1846

27. Postmus, C., Ferraro, J.R., Mitra, S.S.: Pressure dependence of infrared eigenfrequencies of KCl and KBr. Phys. Rev. **174**, 983–987 (1968). https://doi.org/10.1103/PhysRev.174.983
28. Xu, J., Wang, K., Zu, S.Z., Han, B.H., Wei, Z.: Hierarchical nanocomposites of polyaniline nanowire arrays on graphene oxide sheets with synergistic effect for energy storage. ACS Nano **4**, 5019–5026 (2010). https://doi.org/10.1021/nn1006539
29. Ren, X., Li, J., Tan, X., Wang, X.: Comparative study of graphene oxide, activated carbon and carbon nanotubes as adsorbents for copper decontamination. Dalton. Trans. **42**, 5266–5274 (2013). https://doi.org/10.1039/c3dt32969k
30. Tarcan, R., Todor-Boer, O., Petrovai, I., Leordean, C., Astilean, S., Botiz, I.: Reduced graphene oxide today. J. Mater. Chem. C **8**, 1198–1224 (2020). https://doi.org/10.1039/c9tc04916a
31. Bai, X., Wan, C., Zhang, Y., Zhai, Y.: Reinforcement of hydrogenated carboxylated nitrile-butadiene rubber with exfoliated graphene oxide. Carbon **49**, 1608–1613 (2011). https://doi.org/10.1016/j.carbon.2010.12.043
32. Khan, Q.A., Shaur, A., Khan, T.A., Joya, Y.F., Awan, M.S.: Characterization of reduced graphene oxide produced through a modified Hoffman method. Cogent Chemistry, **3**(1–10), 1298980 (2017). https://doi.org/10.1080/23312009.2017.1298980
33. Choi, W., Azad, U.P., Choi, J.P., Lee, D.: Electrocatalytic oxygen reduction by dopant-free, porous graphene aerogel. Electroanalysis **30**, 1472–1478 (2018). https://doi.org/10.1002/elan.201800089
34. Sonkar, P.K., Ganesan, V., Gupta, R., Yadav, D.K., Yadav, M.: Nickel phthalocyanine integrated graphene architecture as bifunctional electrocatalyst for CO_2 and O_2 reductions. J. Electroanal. Chem. **826**, 1–9 (2018). https://doi.org/10.1016/j.jelechem.2018.08.020
35. Wang, Z.J., Weinberg, G., Zhang, Q., Lunkenbein, T., Hoffmann, K.A., Kurnatowska, M., Plodinec, M., Li, Q., Chi, L., Schloeg, R., Willinger, M.G.: Direct observation of graphene growth and associated copper substrate dynamics by in situ scanning electron microscopy. ACS Nano **9**, 1506–1519 (2015). https://doi.org/10.1021/nn5059826
36. Sonkar, P.K., Ganesan, V., Gupta, R., Yadav, D.K.: Dual electrocatalytic behaviour of oxovanadium(IV) salen immobilized carbon materials towards cysteine oxidation and cystine reduction: graphene versus single walled carbon nanotubes. Chem. Select. **1**, 6726–6734 (2016). https://doi.org/10.1002/slct.201601316
37. Folorunso, O., Sadiku, R., Hamam, Y., Ray, S.S.: An investigation of copper oxide-loaded reduced graphene oxide nanocomposite for energy storage applications. Appl. Phys. A **128**, 54–66 (2022). https://doi.org/10.1007/s00339-021-05205-1
38. Nair, R.R., Blake, P., Grigorenko, A.N., Novoselov, K.S., Booth, T.J., Stauber, T., Peres, N.M.R., Geim, A.K.: Fine structure constant defines visual transparency of graphene. Science **320**, 1308 (2008). https://doi.org/10.1126/science.1156965
39. Ni, Z.H., Wang, H.M., Kasim, J., Fan, H.M., Yu, T., Wu, Y.H., Feng, Y.P., Shen, Z.X.: Graphene thickness determination using reflection and contrast spectroscopy. Nano Lett. **7**, 2758–2763 (2007). https://doi.org/10.1021/nl071254m
40. Blake, P., Hill, E.W., Neto, A.H.C., Novoselov, K.S., Jiang, D., Yang, R., Booth, T.J., Geim, A.K.: Making graphene visible. Appl. Phys. Lett. **91**, 063124–063128 (2007). https://doi.org/10.1063/1.2768624
41. Jia, C., Jiang, J., Gan, L., Guo, X.: Direct optical characterization of graphene growth and domains on growth substrates. Sci. Rep. **2**, 707–713 (2012). https://doi.org/10.1038/srep00707
42. Cheng, Y., Song, Y., Zhao, D., Zhang, X., Yin, S., Wang, P., Wang, M., Xia, Y., Maruyama, S., Zhao, P., Wang, H.: Direct identification of multilayer graphene stacks on copper by optical microscopy. Chem. Mater. **28**, 2165–2171 (2016). https://doi.org/10.1021/acs.chemmater.6b00053
43. Korkh, Y.V., Rinkevich, A.B., Klepikova, A.S., Tolmacheva, E.A., Pankrushina, E.A.: Characterization of CVD-synthesized graphene films transferred on different substrates using the scanning probe microscopy electrical techniques. Phys. E **138**, 115101–115112 (2022). https://doi.org/10.1016/j.physe.2021.115101
44. Mattevi, C., Kim, H., Chhowalla, M.: A review of chemical vapour deposition of graphene on copper. J. Mater. Chem. **21**, 3324–3334 (2011). https://doi.org/10.1039/c0jm02126a

45. Lin, Y.M., Dimitrakopoulos, C., Jenkins, K.A., Farmer, D.B., Chiu, H.Y., Grill, A., Avouris, P.: 100-GHz Transistors from wafer-scale epitaxial graphene. Science **327**, 662–664 (2010). https://doi.org/10.1126/science.1184289
46. Wu, M.C., Xu, Q., Goodman, D.W.: Investigation of graphitic overlayers formed from methane decomposition on Ru(0001) and catalysts with scanning tunneling microscopy and high resolution electron energy loss spectroscopy. J. Phys. Chem. **98**, 5104–5110 (1994). https://doi.org/10.1021/j100070a027
47. Klink, C., Stensgaard, I., Besenbacher, F., Laegsgaard, E.: An STM study of carbon-induced structures on Ni(111): evidence for a carbidic-phase clock reconstruction. Surf. Sci. **342**, 250–260 (1995). https://doi.org/10.1016/0039-6028(95)00697-4
48. Land, T.A., Michely, T., Behm, R.J., Hemminger, J.C., Comsa, G.: STM investigation of single layer graphite structures produced on Pt(111) by hydrocarbon decomposition. Surf. Sci. **264**, 261–270 (1992). https://doi.org/10.1016/0039-6028(92)90183-7
49. Binnig, G., Rohrer, H., Gerber, C., Weibel, E.: Surface studies by scanning tunneling microscopy. Phys. Rev. Lett. **49**, 57–61 (1982). https://doi.org/10.1103/PhysRevLett.49.57
50. Fisher, A.J.: Scanning probe microscopy, theory. Encycl. Spectrosc. Spectrom. 2060–2066 (1999). https://doi.org/10.1006/rwsp.2000.0272
51. Fischer, O., Kugler, M., Aprile, M.I., Berthod, C., Renner, C.: Scanning tunneling spectroscopy of high-temperature superconductors. Rev. Mod. Phys. **79**, 353–419 (2007). https://doi.org/10.1103/RevModPhys.79.353
52. Deshpande, A., Bao, W., Miao, F., Lau, C.N., Roy, L.B.J.: Spatially resolved spectroscopy of monolayer graphene on SiO_2. Phys. Rev. B. **79**, 205411–205419 (2009). https://doi.org/10.1103/PhysRevB.79.205411
53. Coraux, J., Diaye, A.T.N., Busse, C., Michely, T.: Structural coherency of graphene on Ir(111). Nano Lett. **8**, 565–570 (2008). https://doi.org/10.1021/nl0728874
54. Li, G., Andrei, E.Y.: Observation of Landau levels of Dirac fermions in graphite. Nat. Phys. **3**, 623–627 (2007). https://doi.org/10.1038/nphys653
55. Xue, J., Yamagishi, S.J., Bulmash, D., Jacquod, P., Deshpande, A., Watanabe, K., Taniguchi, T., Herrero, J.P., Roy, L.B.J.: Scanning tunnelling microscopy and spectroscopy of ultra-flat graphene on hexagonal boron nitride. Nat. Mater. **10**, 282–285 (2011). https://doi.org/10.1038/NMAT2968
56. Berger, C., Song, Z., Li, X., Wu, X., Brown, N., Naud, C., Mayou, D., Li, T., Hass, J., Marchenkov, A.N., Conrad, E.H., First, P.N., Heer, W.A.: Electronic confinement and coherence in patterned epitaxial graphene. Science **312**, 1191–1196 (2006). https://doi.org/10.1126/science.1125925
57. Morita, S., Tsukada, S., Mikoshiba, N.: Scanning tunneling microscopy of Kish graphite and highly oriented pyrolytic graphite in air. J. Vac. Sci. Technol. A **6**, 354–365 (1988). https://doi.org/10.1116/1.575412
58. Tian, J., Cao, H., Wu, W., Yu, Q., Chen, Y.P.: Direct imaging of graphene edges: atomic structure and electronic scattering. Nano. Lett. **11**(3663), 3668 (2011). https://doi.org/10.1063/1.3238296
59. Li, G., Luican, A., Andrei, E.Y.: Scanning tunneling spectroscopy of graphene on graphite. Phys. Rev. Lett. **102**, 176804–176808 (2009). https://doi.org/10.1103/PhysRevLett.102.176804
60. Novoselov, K.S., Geim, A.K., Morozov, S.V., Jiang, D., Zhang, Y., Dubonos, S.V., Grigorieva, I.V., Firsov, A.A.: Electric field effect in atomically thin carbon films. Science **306**, 666–669 (2004). https://doi.org/ https://doi.org/10.1126/science.1102896
61. Bao, C., Song, L., Xing, W., Yuan, B., Wilkie, C.A., Huang, J., Guo, Y., Hu, Y.: Preparation of graphene by pressurized oxidation and multiplex reduction and its polymer nanocomposites by masterbatch-based melt blending. J. Mater. Chem. **22**, 6088–6096 (2012). https://doi.org/10.1039/c2jm16203b
62. Li, M., Lu, X., Jiang, J., Gao, L., Gao, J., Jiang, D.: Green modification of graphene dispersion with high nanosheet content, good dispersibility, and long storage stability. RSC Adv. **12**, 6037–6043 (2022). https://doi.org/10.1039/d1ra08520d

63. Bera, S., Ghosh, M., Pal, M., Das, N., Saha, S., Dutta, S.K., Jana, S.: Synthesis, characterization and cytotoxicity of europium incorporated ZnO-graphene nanocomposites on human MCF7 breast cancer cells. RSC Adv. **4**, 37479–37490 (2014). https://doi.org/10.1039/c4ra06243d
64. Jia, X., Delgado, C.J., Terrones, M., Meuniere, V., Dresselhaus, M.S.: Graphene edges: a review of their fabrication and characterization. Nanoscale **3**, 86–95 (2011). https://doi.org/10.1039/c0nr00600a
65. Delgado, C.J., Herrera, R.J.M., Jia, X., Cullen, D.A., Muramatsu, H., Kim, A.Y., Hayashi, T., Ren, Z., Smith, D.J., Okuno, Y., Ohba, T., Kanoh, H., Kaneko, K., Endo, M., Terrones, H., Dresselhaus, M.S., Terrones, M.: Bulk production of a new form of sp^2 carbon: crystalline graphene nanoribbons. Nano Lett. **8**, 2773–2778 (2008). https://doi.org/10.1021/nl801316d
66. Higginbotham, A.L., Kosynkin, D.V., Sinitskii, A., Sun, Z., Tour, J.M.: Lower-defect graphene oxide nanoribbons from multiwalled carbon nanotubes. ACS Nano **4**, 2059–2069 (2010). https://doi.org/10.1021/nn100118m
67. Sun, Z., Yan, Z., Yao, J., Beitler, E., Zhu, Y., Tour, J.M.: Growth of graphene from solid carbon sources. Nature **468**, 549–553 (2010). https://doi.org/10.1038/nature09579
68. Cao, P., Xu, K., Varghese, J.O., Heath, J.R.: Atomic force microscopy characterization of room-temperature adlayers of small organic molecules through graphene templating. J. Am. Chem. Soc. **133**, 2334–2337 (2011). https://doi.org/10.1021/ja108554p
69. Lu, X., Yu, M., Huang, H., Ruoff, R.S.: Tailoring graphite with the goal of achieving single sheets. Nanotechnology **10**, 269–274 (1999). https://doi.org/10.1088/0957-4484/10/3/308
70. Shearer, C.J., Slattery, A.D., Stapleton, A.J., Shapter, J.G., Gibson, C.T.: Accurate thickness measurement of graphene. Nanotechnology **27**, 125704–125715 (2016). https://doi.org/10.1088/0957-4484/27/12/125704
71. Deng, B., Hou, Y., Liu, Y., Khodkov, T., Goossens, S., Tang, J., Wang, Y., Yan, R., Du, Y., Koppens, F.H.L., Wei, X., Zhang, Z., Liu, Z., Peng, H.: Growth of ultraflat graphene with greatly enhanced mechanical properties. Nano Lett. **20**, 6798–6806 (2020). https://doi.org/10.1021/acs.nanolett.0c02785

Potential Applications of Graphene

Asma Mukhtar, Syed Salman Shafqat, Muhammad Nadeem Zafar, Syed Rizwan Shafqat, Mian Habib-Ur-Rahman Mahmood, and Shahid Bashir

Abstract Nanomaterials have revolutionized the fields of science and technology leading towards the growth of innovative materials with improvements. Graphene nanomaterials having unique features such as elevated mechanical power, sharp conductivity, excellent thermal and chemical steadiness, combining efficiency with functionalized groups forming composites have made graphene ideal for its applications in different fields. Graphene and its derivatives including 2D graphene nanosheets, 3D porous graphene, graphene oxide (GO), reduced graphene oxide (rGO) and graphene quantum dots (GQDs) have been encouraged in the fields of energy storage devices, portable and foldable devices, flexible optoelectronics, bioengineering and water purification etc. This chapter includes a general overview on applications of graphene based materials in the fields of energy storage devices, biomedical applications and water purification has been with updated literature.

Keywords Graphene · Energy storage devices · Flexible optoelectronics · Bioengineering · Water purification

A. Mukhtar · S. S. Shafqat (✉) · M. H.-U.-R. Mahmood
Department of Chemistry, Division of Science and Technology, University of Education, Lahore 54600, Pakistan
e-mail: salman.shafqat@ue.edu.pk

M. N. Zafar
Department of Chemistry, University of Gujrat, Gujrat, Pakistan

S. R. Shafqat
Department of Chemistry, University of Sialkot, Sialkot 51310, Pakistan

S. Bashir
Higher Institution Centre of Excellence (HICoE), UM Power Energy Dedicated Advanced Centre (UMPEDAC), Level 4, Wisma R&D, Universiti Malaya, Jalan Pantai Baharu, 59990 Kuala Lumpur, Malaysia

© The Author(s), under exclusive license to Springer Nature Singapore Pte Ltd. 2023
R. T. Subramaniam et al. (eds.), *Graphene*, Engineering Materials,
https://doi.org/10.1007/978-981-99-1206-3_7

1 Introduction

Graphene has surfaced as one of the most promising ***nanomaterials*** due to unique combination of extraordinary properties. The exceptional characteristics of graphene came from the 2p orbitals making it a 2D material. Graphene is stronger with respect to diamond and flexible with respect to rubber, harder with respect to steel but weightless with respect to aluminum, making graphene the outstanding material. Graphene exists in 1D, 2D and 3D structures as shown in Fig. 1 [1].

The distinctive properties of graphene have provided a platform of interesting matrix for upgradation of technologies in different areas—clothes made of electronics, supersensitive sensors, multidisciplinary composite and layering, membranes, drugs and biotechnology, energy scavenging and storage etc. [3, 4].

Based on unique features of graphene, its applications in different areas will be discussed in detail in this section [5].

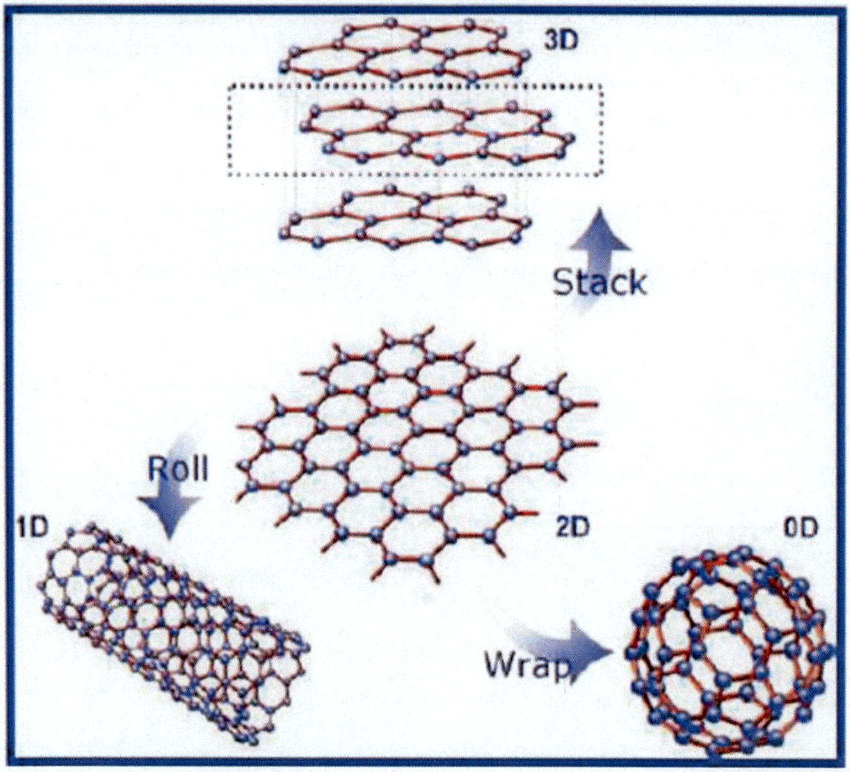

Fig. 1 Schematic diagram illustrates the different forms of graphene. Reproduced with permission Wan et al. [2]

2 Graphene Based Material for Energy Storage Devices

Electrochemically Energy Storage (EES) has a prominent place in our routine life as they have vast applications in electronic devices used in mobile, grid stations, and electric vehicles (EVs) [6]. In the EES systems, batteries that are made of metal ions and electrochemical capacitors are two prominent families [7]. To fulfil the increasing demands at the industrial level and at the community level, advanced EES systems with properties such as higher energy density and better recycling capability are in high demand [8].

Recently, graphene has fascinated attention for its wide application in EES based on its unique physicochemical characteristics [9]. The graphene have characteristics of both metal and nonmetallic materials creating diverse potential specifically in photovoltaic energy storage and energy conversion [10–13]. The extraordinary electrical conductivity, remarkable mechanical properties and large surface area of graphene, physicochemical stability, have proved its likely possibility to initiate the current EES landscape [14–17].

One of the unique features of graphene for its wide applications in the field of energy storage devices is its flexibility which enhanced its variety of different usage and installation conditions [18]. Moreover, flexibility characteristics such as bending, folding or stretching has been used in numerous equipment applications, for example in lithium ion batteries, nanogenerators (NGs), fuel cells and supercapacitors (SCs), respectively [19–22]. Based on the flexibility, graphene exists in four different forms which include graphene paper, graphene film, graphene foam, and graphene fiber. The following Fig. 2 illustrates the diversified potential of graphene applications.

Composites based on graphene, which are formed by combining graphene and electrochemical materials have attracted attention for electrochemical energy storage (EES) devices due to its superior electrochemical performances, which have served the basis for application of these composites in different kinds of EES systems. The graphene component in composites has introduced new characteristics to composites such as elasticity, translucence, self-recovering properties, etc. Typically the composites of graphene include composites made of carbon, graphene–metal, graphene–metal oxide and graphene–polymer composites [23–25]. The graphene based composite materials formed comprise of inorganic nanoparticles and polymers along with graphene. For instance, graphene-polyaniline (PANI) composites have been used in the form of paper sheet for conductivity. This paper of graphene-PANI composite gives high success for development of low-cost electrode in energy storage devices [26–28].

Interestingly, the graphene structures also come in forms of 2D and 3D structures, respectively. The 3D graphene structures have distinctive properties such as electrical conductivity, extraordinary structural stability, and fast diffusion channels for ions. 3D graphene structure includes graphene aerogels, graphene foams (GFs), graphene networks, and so on. One of graphene 3D structure such as graphene-PLA (polylactic Acid) filament has been used as energy storage architectures [29]. Furthermore, the

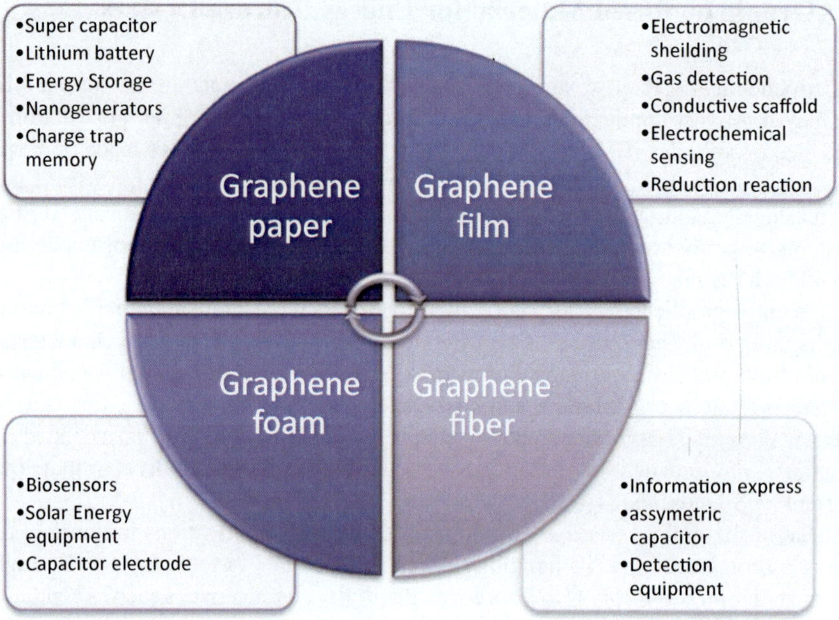

Fig. 2 Different forms of graphene and its applications

3D graphene that absorbs the metal oxide forming 3D composites have exhibited harmonious properties and multiple functionalities [30–33].

In order to increase the electrochemical efficiency of composites made from graphene, the porous graphene structures are fabricated by means of developing holes inside the graphene sheets and by fabricating the sheets in a 3D porous framework. Reduced form of graphene oxide imparts quick ion diffusion with large surface area [34]. In addition, the graphene composite, porous graphene, 2D graphene, 3D graphene, GOs have been extensively used in energy storage devices and will be discussed in the upcoming sections.

3 Graphene Based Material for Batteries

From the past two decades, life has become very electronic by using electronic devices such as laptops, mobile phones, and electric automobiles. Increased use of these devices has increased the interest in developing efficient materials for energy storage having great power, high energy concentration and fast rate of charging. Furthermore, batteries and supercapacitors represent the most common electrochemical energy storage devices [35–38].

3.1 Li-Ion Batteries

Li-ion batteries (LIBs), the most demanding types of EES devices, have unique features such as high energy density, elongated recycling life, and reduced self-discharging have gained attention from both the scientific society and community. In LIBs, cathode releases lithium and is captured by anode i.e., graphite, in which Li^+ ions continuously move back and forth. The efficiency of any battery in terms of energy and gravimetric capacity is determined by the quantity of ions carried in one gram of product [39].

In general, there are few flaws such as Li^+ ion insertion in graphene showing a high irreparable capability and the unpreventable re-assembling of graphene lowers the Li^+ ion storage ability processes. When a monolayer of graphene is used, the specific capacity of graphene doubles up as compared to graphite. Based on this observation, replacing graphite with graphene as anode for capturing Li^+ ion directly or by forming a composite of graphene by combining it with other Li-hosting components has made batteries more advanced and efficient [39].

The composites used as active components include metal oxides and metals. These composites can be developed by reversible insertion reaction of TiO_2, by making alloys of stannic oxide (SnO_2), tin (Sn) or by conversion reaction of Iron (II, III) oxide Fe_3O_4, Cobalt (II, III) oxide Co_3O_4, etc. with Li^+ ion thus providing considerable enhanced capacities. Runwei Mo et al. explored 3D interlinked porous graphene foam doped in nitrogen encapsulated with Ge quantum dot for LIB anode which exhibited high reversible specific capacity, long cycling capability (over 96%) and ultra-high rate performance [40]. Xiangfeng Duan's group developed a composite of 3D holey-graphene/Nb_2O_5 which were maximized to convey ultrahigh-rate energy storage performance [41]. Additionally some perfectly transformable capabilities are also obtained for silicon/graphene, Fe_3O_4/graphene and Co_3O_4/graphene composites [39].

Lithium cobalt oxide ($LiCoO_2$), lithium manganese oxide ($LiMn_2O_4$) and Lithium iron phosphate battery ($LiFePO_4$) are mostly explored cathodic materials for LIBs but due to reduced electronic conductivity, reduced rate capability these cathode materials were sharpening for their efficiencies by incorporating the fabricated lithium with 3D graphene. 3D graphene networks and graphene-based membranes have been studied as the alternative anodes of LIBs having unique features such as large specific surface area, a constant electron-transport pathways and rapid ion-diffusion, extraordinary mechanical strength making a more promising energy storage device.

Duan and his coworkers developed a 3D solvated graphene framework (SGF) which showed a high reversible capacity, excellent rate and cycling ability [42]. Kang et al. developed a cathode of V_2O_5/graphene membrane and an anode of lithiated graphene membrane. The graphene membrane in cathode served as a conducting agent as well as a current collector. The assembled battery showed a capacity of 15 mAh/cm^2 [43]. Li et al. observed that nanosheets of Nickle oxide nanostructures (NiO NSs) bonds to graphene enhanced the performance of the cathode in terms of reversible capacity and rate performance. Moreover, in another investigation, it was revealed that the adsorption of Li atom to TiO_2 surface was increased significantly due to the graphene modification [44].

Alternatively, when pores were introduced into the 3D graphene framework (sheets), its effectiveness increased for transporting ion diffusion channels. Wu et al. prepared a porous 3D graphene network encapsulated with Sn-based nanoparticles that served as LIB anode. This anode showed remarkable efficiency in transferring electron and high diffusion of Li^+ ions into the Cobalt tin oxide ($CoSnO_3$) particles incorporated into the porous 3D conductible framework Sun et al. developed a nanocomposite of 3D holey-graphene/niobia (Nb_2O_5) where 3D graphene was used as the conductive material and Nb_2O_5 were used as nanoparticles [29].

3.2 Lithium–Sulfur Batteries

In comparison to LIBs, Li–S batteries uses anodic lithium and cathodic sulfur [45]. The high energy density of Li–S battery and low cost of sulfur has attained high attention and high improvements in recent years. Although, there are few problems under consideration which include reduced electrical conduction and high-volume difference of sulfur occurred during charging/discharging of battery. One of the solutions is to use graphene-sulfur hybrid having huge surface area, high level conduction and superior elasticity properties [46]. Dai et al. synthesized a graphene-sulfur composite having a three dimensional "sandwich-like" structure [47]. Zhang et al. explored the encapsulation sulfur by using a double-layered graphene film. The elasticity of graphene helped to wrap or encapsulate the nanoparticles which results to design a core-shelled figure with high discharge capacities. However, to improve the sulfur cathode efficiency, Manthiram et al. explored a double refinement method where the carbon sphere was doped with N and was encapsulated by graphene [48]. Cheng et al. reported a fibrous hybrid made up of graphene-sulfur having extraordinary electrochemical efficiency with an increased specific capacitance. In order to enhance the Li–S batteries efficiency, Yang et al. prepared sulfur-graphene (S-G) nano-walls which was effective to enable the rapid diffusion of electrons and Li ions [44]. Gu et al. designed a nitrogen and phosphorous doped porous and conductive graphene. Here, the results showed the conductivity of co-doped graphene to be higher than that of nitrogen- or phosphorous-doped graphene layer [49].

In Li–S batteries, the formed polysulfides were failed to be trapped by graphene. 3D graphene having porous morphology has been used to trap polysulfide species [50]. Xu et al. prepared a cathode material made of 3D graphene sulfur composite with metal ions [51]. The resulted material consisted of 3D nano sized porous sheets having pore size in the range of 20–200 nm. The porous surface prevented the dissociation of polysulfides, hence, enhancing the battery's efficiency.

Porous graphene being hierarchical has pores of different sizes with different functionalities. Hierarchical graphene having macropores functions as mass-controlling areas which help in easy diffusion whereas mesopores reduce the path for movement of ions and micropores increases the surface area which provides large active sites

[52]. Tang et al. developed a hierarchical porous graphene which served as a cathode with high capacity. A hybrid carbon aerogel (HCA) film exhibited a high reversible capability as well as a high-rate capacity in Li–S batteries. A cathode made-up of graphene aerogel and TiO_2 composite loaded with S was used for Li–S batteries [53–55].

In order to avoid the polysulfides to get dissolved, 3D graphene has been doped with a variety of heteroatoms such as N, S, B, and F and loaded with S to prepare an electrode for Li–S batteries. Zhou et al. prepared a negative electrode of Li_2S encapsulated with 3D graphene aerogels doped in B and N for Li–S batteries as depicted in Fig. 3 [56, 57].

3.3 *Na-Ion Batteries*

Based on the availability of large reserves of sodium, the sodium-ion batteries (SIBs) are considered an alternative to LIB. Additionally, the characteristics such as having high voltage, efficient charge–discharge recyclability, and safety have made SIB a promising alternative to LIB. Porous graphene exists as in plane and out of the plane and in both SIB and LIB in plane porous graphene are used. Luo and his coworkers studied the porous graphene nanosheets having functional groups adsorbed on the surface affecting the storage of Na^+ ion. The electrodes made up of hybrid of nanosheets of holey graphene mixed up with a huge quantity of functional groups were developed which showed a sharp reversible capability [58, 59].

Rui et al. prepared a spongy graphene framework encapsulated with sodium vinyl phosphate ($Na_3V_2(PO_4)_3$). The graphene electrode composite made up of reduced graphene oxide mixed up with carbon nanotubes and sodium vinyl phosphate ($Na_3V_2(PO_4)_3$) showed a sharp electrical conduction and quick movement of positive and negative ions; this results in completion of a cycle of charging discharging within 1 min [59].

Zhao et al. developed a positively charged electrode for Na^+ ion batteries by partial reduction of perforated GO upon hydrogen peroxide H_2O_2 treatment. The positively charged electrode for sodium-ion batteries showed high rate capability and excellent cycling performance representing the efficient actions for sodium-ion batteries made of graphene and its derivatives [60].

Yang and his coworkers prepared graphene-based composite doped with nitrogen encapsulated with nickel cobaltite (NCO) which exhibited efficient electrochemical reactions in sodium-ion batteries. The graphene based composites enhanced the rate capabilities with elevated specific capacitance and a good cycling stability [61].

Graphene composites have been used as anode by using active materials such as oxide of transition metals (e.g., SnO_2, Fe_3O_4, $MgFe_2O_4$), sulfides of transition metals (e.g., SnS_2, MoS_2), metals or metal alloys (e.g., Sb, P), phosphorene, and NASICON-type $NaTi_2(PO_4)_3$ for NIBs [62–70]. Yi Cui's group prepared a graphene composite made-up of 2D phosphorous (phosphorene) developed by layering the 2D

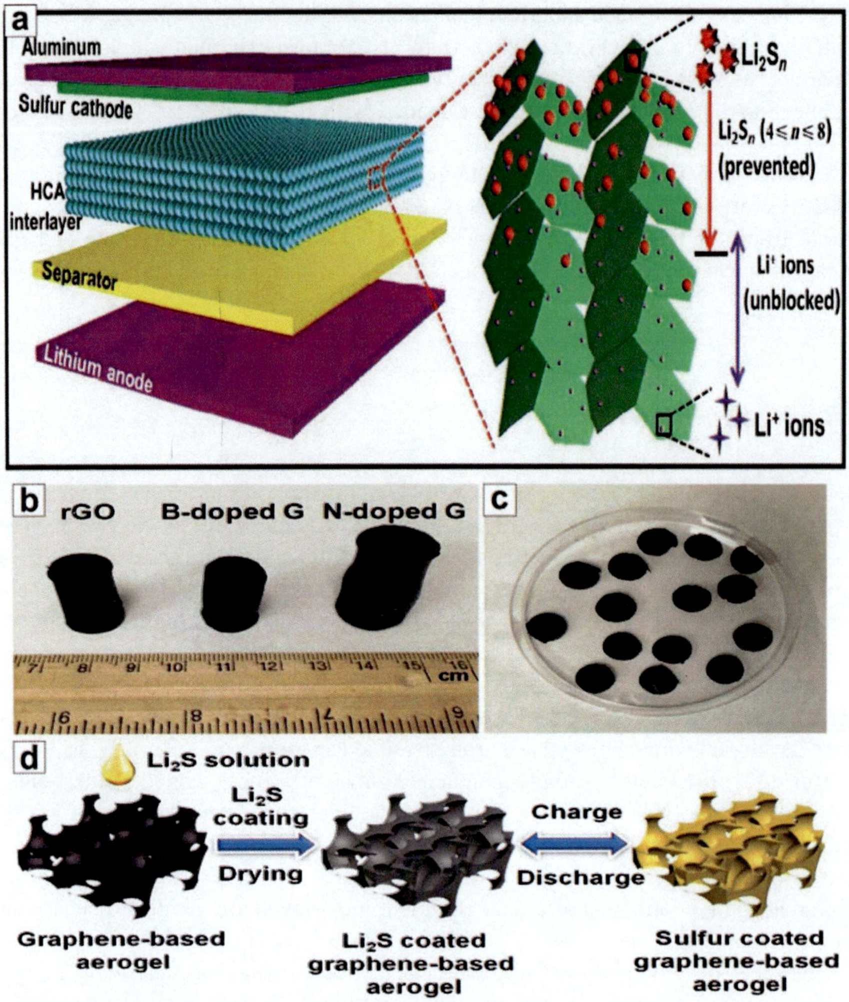

Fig. 3 Schematic model illustrates the **a** detailed structural features of a Li–S battery, **b** N doped and B doped graphene and GO. **c** Small pieces of B-doped graphene hydrogel and **d** step by step coating of Li_2S respectively. Reproduced with permission from Zhou et al. [57]

phosphorous in graphene layers showing a high specific capacity [69]. The layering of graphene and composite provided pathways for electron movement [61].

In general, the graphene component is an inactive cathodic material for SIB, but when composite based graphene (combined graphene with FeF_3, $Na_3V_2(PO_4)_3$, $Na_2/3Fe_1/2Mn_1/2O_2$ etc.) is designed, a high electronic conductivity with improved rate performance was observed [59, 71].

Yunhui Huang's group prepared a composite of 3D graphene encapsulated with $NaTi_2(PO_4)_3$ which enhanced the conductivity, ionic character, and electro mechanical properties of the electrode with high reversible capacity [43]. Yu Zhang's group prepared an excellent cathode developed by a hybrid of reduced graphene oxide with FeF_3-Fe having superior discharge capability, good cycling stability and elevated-power efficiency [39].

3.4 *Li-Air Batteries*

Li air batteries have capacity to store additional energy in comparison to LIBs (lithium-ion batteries). When Li air batteries operate the Lithium reduces oxygen and forms LiO_2. This lithium oxide blocks the electrolytes and passage of the oxygen which reduces the cycle lifetime, poor rate capacities and hence poor efficiency of the battery. For this purpose, graphene formed a composite with porous nanostructures of noble gas, transition metals or oxides of transition metals such as Pt, Pb, Co_3O_4, $MnCo_2O_4$, with characteristics such as high surface area. For instance, Shui et al. prepared holey graphene doped with nitrogen, a hierarchical porous cathode, which was effective for the movement of oxygen, electrolytic solution, and electron. Such electrode exhibits superb electrochemical properties such as excellent gravimetric capability. Moreover, different batteries based on graphene and their efficiency are described in Table 1 [72, 73].

4 Graphene Based Material for Supercapacitors

In comparison to EES devices, the electrochemical capacitors (ECs) have high energy density with a sharp charge-storage mechanism. These ECs are also referred to as supercapacitors or ultracapacitors. The double layered electrochemical devices due to graphene are a future perspective. Lots of research work has been performed utilizing graphene in different forms and its capacitance has been recorded in the range of 100 and 200 F/g [82]. Furthermore, previous conducted studies explored that the incorporation of graphene into a supercapacitor is responsible to enhance the surface area of graphene, functioning as electrode, and its conductivity [83]. Graphene has been used widely as electrodes of supercapacitors. As mentioned previously, graphene has different structures such as 1D (fibers and thread types), 2D (graphene and graphene-based films), and 3D (foams, aerogels, and sponges of graphene) which are used as supercapacitors. Graphene also exists in porous form known as holey graphene that shows superior performance as compared to non-porous graphene sheets [84–86].

2D graphene such as graphene sheets have unique characteristics including lightweight, larger surface area, sharp electrical conduction, variable thickness and

Table 1 Electrochemical performances of the batteries of Li-ion, Li–S and Na-ion batteries

Material	Type of porosity	Capability to discharge at initially	Rating ability	References
Lithium-ion battery				
$CoSnO_3$/framework of porous graphene	Away the plane	At 1 A g^{-1} the initial discharge capacity is 1380 mA h g^{-1}	At 10 A g^{-1} the rate is 432 mA h g^{-1}	[74]
MoS_2intermixed with 3D graphene	Away the plane	At 0.5 A g^{-1} the initial discharge capacity is 1403 mA h g^{-1}	At 2 A g^{-1} the rate is 780 mA h g^{-1}	[75]
Composite made of SnO_2 with mesoporous graphene	Away the plane	At 0.6 A g^{-1} the initial discharge capacity is 1807 mA h g^{-1}	At 15 A g^{-1} the rate is 425 mA h g^{-1}	[76]
Graphene encapsulated with Iron oxide- and gelatin	Away the plane Within the plane	At 0.1 A g^{-1} the initial discharge capacity is 2518 mA h g^{-1}	At 0.8 A g^{-1} the rate is 634 mA h g^{-1}	[77]
Lithium–sulfur battery				
Porous graphene doped in nitrogen and phosphorous	Within the plane	At 1 C the initial discharge capacity is 1158.3 mA h g^{-1}	At 2 C the rate is 633.7 mA h g^{-1}	[49]
Composite of S3D-graphene	Away the plane	At 0.1 C the initial discharge capacity is 1260 mA h g^{-1}	At 2 C the rate is 500 mA h g^{-1}	[51]
Hierarchical porous graphene	Within the plane	At 0.5 C the initial discharge capacity is 1053 mA h g^{-1}	At 5 C the rate is 357 mA h g^{-1}	[53]
Porous graphene network	Away the plane	At 1 C the initial discharge capacity is 1187 mA h g^{-1}	At 10 C the rate is 980 mA h g^{-1}	[78]
Hybrid of 3D graphene-foam with reduced graphene oxide^{-}	Away the plane	At 0.2 C the initial discharge capacity is 1000 mA h g^{-1}	At 2 C the rate is 538 mA h g^{-1}	[79]
Composite of carbon nanotube with 3D graphene	Away the plane	At 0.2 C the initial discharge capacity is 1179.6 mA h g^{-1}	At 4 C the rate is 653.4 mA h g^{-1}	[80]
Sodium-ion battery				
Nanosheets Porous graphene	Within the plane	At 0.03 A g^{-1} the initial discharge capacity is 220 mA h g^{-1}	At 10 A g^{-1} the rate is 85 mA h g^{-1}	[58]

(continued)

Table 1 (continued)

Material	Type of porosity	Capability to discharge at initially	Rating ability	References
3D porous Sb_2S_5-graphene foam	Away the plane	At 0.1 A g^{-1} the initial discharge capacity is 1156 mA h g^{-1}	At 10 A g^{-1} the rate is 525 mA h g^{-1}	[81]

effective mechanical property making it a superior material for being a supercapacitor. For instance, the supercapacitor prepared by graphene papers layered with nanoparticles of black carbon or CNTs improved the energy storage capability. 3D graphene which includes graphene foams, sponges, and aerogels were studied for supercapacitor applications based on the porous structure, large surface area and fast electron movement channels. The 3D graphene composite, for example, MnO_2-graphene foam forming the hybrid electrode achieved a capacitance of 389 F/g with a compactness of 25 kW/kg and energy density of 44 Wh/kg [87]. The 3D composite structure NiO-graphene, Co_3O_4 nanowires-graphene foam showed elevated specific capacitance. A graphene composite made of 3D graphene with polymers such as PVA, multi-redox anthraquinone derivative alizarin and polyhydroquinone achieved elevated energy density [86].

Sui et al. prepared N doped porous graphene which reported that these doped graphene electrodes further enhanced the wettability and electrical conductivity which improves the overall performance of the supercapacitor. To date, a variety of different 3D porous graphene electrodes have been developed having high volumetric density. Xu et al. prepared a 3D holey graphene framework (HGF) having a hierarchical porous structure as an electrode with sharp-energy-capacitance for the supercapacitor [88]. Qu et al. developed a graphene composite made up of hierarchical graphene aerogel encapsulated with polyaniline (PANI), (HGC-PANI) which resulted in increased capacity and cycling rate [34].

The performance of the supercapacitor of different 3D graphenes and its derivatives has been compiled in Table 2. It can be concluded that 3D graphene has provided high supercapacitor effectiveness based on the efficient conductivity and penetrable structure. By preparing the electrode of the 3D graphene-based composite, an improved electrochemical performance of supercapacitors was observed. The improvement of efficiency of electrodes of the supercapacitors was due to both the 3D graphene and active material. By heteroatom doping, the 3D graphene further enhanced the performance of supercapacitors in terms of enhancing the wettability and conductivity [34].

On the basis of capacitance of the capacitor, supercapacitors developed by graphene are classified into two major groups: electrochemical double layer capacitors and redox based electrochemical capacitor.

Table 2 Graphene and their composites used in supercapacitors

Material	Type of porousness	Precise capacitance	Cycling sustainability	Rating ability	References
Electrochemical double layer capacitor					
Holey graphene hydrogels	Within the plane Away the -plane	At 1 A g^{-1} the capacitance is 283 F g^{-1}	20 000 cycles at 20 A g^{-1} showed cycling stability of = 94%	At 100 A g^{-1} the rate is = 75%	[89]
Multilayerd porous graphene paper	Within the plane	At 1 A g^{-1} the capacitance is 209 F g^{-1}	20 000 cycles at 10 A g^{-1} showed cycling stability of ≈90%	At 20 A g^{-1} the rate is 75%	[89]
Holey graphene	Within the plane	At 1 A g^{-1} the capacitance is 190 F g^{-1}	6000 cycles at 5 A g^{-1} showed cycling stability of 94%	At 60 A g^{-1} the rate is 78%	[90]
Nanomesh sized graphene	Within the plane	At 1 A g^{-1} the capacitance is 253 F g^{-1}	50 000 cycles showed cycling stability of 91%	At 100 A g^{-1} the rate is 84%	[91]
Disk of holey graphene	Within the plane	At 1 A g^{-1} the capacitance is 30 F g^{-1}	10 000 cycles showed cycling stability of 80%	–	[90]
Nitrogen and sulfur doped graphene nanosheets	Within the plane away the -plane	At 0.8 A g^{-1} the capacitance is 146.6 F g^{-1}	25 000 cycles at 3.2 A g^{-1} showed cycling stability of 94.5%	At 3.2 A g^{-1} the rate is ≈80%	[92]
Hydrogel doped in nitrogen	Away the -plane	At 1 A g^{-1} the capacitance is 255 F g^{-1}	2000 cycles showed cycling stability of 100%	At 20 A g^{-1} the rate is 82%	[93]

(continued)

Table 2 (continued)

Material	Type of porousness	Precise capacitance	Cycling sustainability	Rating ability	References
Polyurethane sponge derived reduced graphene oxide	Within the plane away the -plane	At 1 A g^{-1} the capacitance is 186 F g^{-1}	4000 cycles showed cycling stability of 93%	At 20 A g^{-1} the rate is ≈45%	[94]
Redox-based capacitor					
Composite of hydrogel of graphene with MnO_2	Away the -plane	At 0.5 A g^{-1} the capacitance is 321.4 F g^{-1}	5000 cycles showed cycling stability of 82.4%	At 8 A g^{-1} the rate is 57.9%	[95]
3D graphene encapsulated MnO_2	Away the -plane	At 0.3 A g^{-1} the capacitance is 659.7 F g^{-1}	1000 cycles showed cycling stability of 106%	At 10 A g^{-1} the rate is ≈53%	[96]
Composite of CNTwith aerogel of 3D graphene	Away the -plane	At 1 mA cm^{-2} the capacitance is 8.56 F cm^{-3}	20 000 cycles showed cycling stability of 84.6%	At 20 mA cm^{-2} the rate is ≈58%	[97]
Composite of Polyaniline nanorod with nanomesh of graphene	Within the plane away the -plane	At 1 A g^{-1} the capacitance is 452 F g^{-1}	2000 cycles at 3.0 A g^{-1} showed cycling stability of 77%	At 50 A g^{-1} the rate is 84%	[98]
Composite of aerogel graphene with polyaniline and Co_3O_4	Away the -plane	At 1 A g^{-1} the capacitance is 1247 F g^{-1}	3500 cycles showed cycling stability of 100%	At 20 A g^{-1} the rate is 60.5%	[98]
Ppy nitrogen phosphorous doped aerogel of porous graphene	Within the plane away the -plane	At 0.5 A g^{-1} the capacitance is 418 F g^{-1}	2000 cycles showed cycling stability of 74%	At 20 A g^{-1} the rate is 76.9%	[99]
Composite of NiCo–LDH with reduced graphene oxide	Away the -plane	At 2.0 A g^{-1} the capacitance is 1911.1 F g^{-1}	1000 cycles at 20 A g^{-1} showed cycling stability of 74%	At 20 A g^{-1} the rate is 76.9%	[100]

4.1 Electrochemical Double Layer Capacitors (EDLCs)

In EDLCs, the charge stored is proportional to the double layer contact with the electrolyte. Graphene and its modified form served as electrodes of EDLC due to characteristics such as large surface area, elevated electrical conductivity, and successful charge transfer rate hence having elevated power capacities. Usually, graphene having surface area of 2630 $m^2 g^{-1}$ can have capacitance up to 550 F g^{-1}, showing a sharp potential for developing high energy density of such systems [101]. Xu and his coworkers used the holey graphene framework (HGF) which has a lot of in-plane nanopores with an elevated gravimetric capacitance of 298 F g^{-1} at 1 A g^{-1} using an electrolyte (organic in nature) [39].

Recently, a binder free electrode of holey graphene hydrogel (BNP-HGH) doped in nitrogen, boron, and phosphorus was utilized to attain elevated energy capacities supercapacitors.

A variety of different graphenes have been used such as activated graphene, 3D architectural graphene, graphene based composites, by varying the orientation of graphene which becomes 90° to the current collector plane, hence, enhancing the capacitors capacity [39].

Composites based on graphene have also been prepared for enhancing the infiltration and increasing the electrical conductivity of the electrode. These composites were prepared by coating graphene with different forms of carbon e.g., fullerene, black carbon, nanotubes of carbon, porous and hollow carbon spheres, activated carbon, etc. [84–86, 102, 103].

Zhuangjun Fan's group prepared a graphene/carbon black composite which acted as a conducting material as well as a spacer in layering of the graphene sheets which increased electron and ion transportation [85]. Similarly, Xue Qin's group reported the graphene/fullerene composite where fullerene was used as the spacer and attained a higher capacity of 135 $F{\cdot}g^{-1}$ at the current density of 1 $A{\cdot}m^{-2}$. For the fabrication of graphene based composites, carbon nanotubes (CNTs) of one dimension and carbon spheres of three dimension and activated carbon have also been used which enhanced the EDLC efficiency [39].

Liming Dai's group prepared graphene/CNTs composites by layering graphene with CNTs having an average specific capacity of 120 F g^{-1} [104]. Dengsong Zhang and Liyi Shi's group also developed the composite of graphene/CNTs with a much higher specific capacitance (230 F g^{-1}) than graphene (101 F g^{-1}) [105]. Along with 1D, 3D graphene/carbon nanotube composite was also fabricated resulting in sharp capacitance. For instance, Likun Pan's group reported graphene based composite prepared by 3D activated carbon with graphene which increased the specific capacitance of 181 F g^{-1} [103].

4.2 Redox Based Electrochemical Capacitors

EDLCs transfer energy is based on the energy storage mechanism while pseudocapacitors store energy at a reduced power. Pseudocapacitors are also known as redox based electrochemical capacitors. One of the drawbacks of capacitors is the low energy density. To overcome the effect, hybrid electrodes can be prepared by encapsulating the oxides of metal, hydroxides of metal or conductive polymers with graphene. The graphene serves as a conductive pathway for transferring charge. Graphene based composites may also include carbon species, polymers of polyaniline, polypyrrole and polythiophene which showed conductivity, oxides of metal (e.g., Mn_3O_4, Co_3O_4, SnO_2 etc.), metals, or nitrides (e.g., VN, Ni_3N etc.), sulfides (e.g., FeS_2, MoS_2, etc.), hydroxides (MnNi-LDH, NiCo-LDH etc.) and MXenes etc. Graphene-MnO_2 composites were used as electrodes for the supercapacitor and it was observed that the specific capacitance of 211.5 F/g at 2 mv/s while using 1 M Na_2SO_4 electrolyte was achieved and 75% of the capacitance was retained even after 1000 charge–discharge cycles [39].

When the graphene-Ni $(OH)_2$ composite was used as a hybrid electrode, a high specific capacitance of 953 F/g and high power and high energy densities were obtained. It was observed that the Fe_3O_4-rGO composite showed more than double sharper specific capacitance when compared to Fe_2O_3. Similarly, the graphene composite, Co_3O_4-rGO, also exhibited a superior rate capability and cycling capacity [39].

Recently, Yury Gogotsi's group prepared a composite of graphene with 2,5-dimethoxy-1,4-benzoquinone ((DMQ)/RGO) [106]. Jun Yan et al. prepared a composite of the graphene film with MXene having volumetric capacitance of 1040 F cm^{-3} at 2 mV s^{-1} with efficient rate ability and long cycle life [107]. Alternatively, Songhun Yoon's group developed a 3D graphene composite of MnNi-LDH where MnNi-LDH was incorporated between layers of 2D graphene nanosheets. The graphene composite retained a sustainable efficiency up to 90% capacity reservation after 1400 rotations and a sharp energy capacity of 47.29 kWh [39].

5 Graphene-Based Material for Flexible Optoelectronics

Optoelectronics are electronics which transform the electrical energy into light. Optoelectronic devices including photodetectors, modulators, lasers, light emitting diode (LED) and solar cells, etc. that are involved in different applications such as imaging, display, optical communications, healthcare, security, and energy stacking. Graphene has attained a lot of demand in the field of optoelectronics based on its extraordinary electronic and optical properties including super-high carrier motility, zero bandgap, tunable work function, fast carrier dynamics and strong light-matter coupling [108–111].

Table 3 Graphene electrodes in organic solar cell and dye-sensitized solar cells

Devices	Medium	Electrode	PCE [%]	References
Organic solar cells (OSCs)	CVD	Anode placed at Base	2.5	[112]
	Reduced graphene oxide	Anode placed at Base	0.78	[113]
	CVD	Anode placed at Base	2.6	[114]
	CVD	Anode placed at Base	5.3	[115]
	CVD	Anode placed at Base	0.84	[116]
	CVD	Anode placed at Base	2.43	[117]
	CVD	Anode placed at Base	4.33	[118]
	CVD	Anode placed at Base	2.07	[119]
	CVD	Base cathode	3.13	[120]
	CVD	Anode placed at Base	1.18	[121]
	CVD	Anode placed at Base	7.51	[122]
Dye-sensitized solar cells (DSSCs)	Graphene sheets	Counter	5.73	[123]
	Nitrogen doped-GF	Counter	7.07	[124]
	Reduced graphene oxide	+Electrode	0.26	[125]
	Graphene sheets	Counter	4.99	[126]

Due to unique properties of graphene, the commercially available indium tin oxide (ITO) has been replaced by graphene as glassy conductive electrodes for optoelectronic devices. The efficiencies due to graphene electrode of various optoelectronic devices are illustrated Table 3.

Graphene electrodes have been used as bottom Transparent conducting electrodes (TCEs) and top TCEs. Graphene electrodes as bottom TCEs have been used in organic solar cells. Many scientists globally have participated using graphene as transparent conductive electrodes in organic solar cells. The rGO has been used as the bottom electrode in organic solar cells with formulation of device of reduced graphene oxide with PEDOT:PSS/P_3HT:PCBM/TiO_2/Al exhibited PCE, a power conversion efficiency of 0.78%. By altering the layer of graphene and enhancing the intrinsic transparency and conductivity of it the PCE also increases. Arco et al. utilized graphene as an anode in OSCs prepared by the CVD method [120, 121].

Jung et al. prepared graphene electrode doped in nanoparticles of zinc oxide (ZnO). A n-type dopant has been used for graphene cathode in organic solar cells [127]. Yoon et al. prepared elastic PSCs based on graphene transparent conductive electrodes that attained power conversion efficiency of 16.8% without any hysteresis [128]. Dye-Sensitized Solar Cells (DSSCs) as photovoltaic devices have also been explored because of their sharpness in energy conversion and consumption of low-cost for production. Roy-Mayhew and his coworkers used catalytic electrodes made up of composite of graphene in DSSCs, which showed the efficiency of 5.0% [126].

To enhance the effectiveness, the nitrogen and sulfur dopped graphene (NSG) were also prepared [129].

Graphene-based photodetectors are efficient because their response toward a broad spectrum ranging from visible to infrared region is sharp. The absorption of light can be enhanced by incorporating quantum dots (QDs) on the graphene surface. When doped graphene was coupled with QDs, the charge transfer was enhanced compared to simple graphene coupled with QD. Exfoliated graphene encapsulated with PbS QDs provided a very high photoresponsivity. Peng et al. explored composites of flexible multiwalled carbon nanotube (MWCNT)/graphene hybrid materials incorporated with cadmium telluride (CdTe) QDs. The resulting electrodes exhibited a sharp reversible photocurrent [130].

6 Graphene-Based Material for Touch-Based Flexible Screens and OLED Displays

Currently, ultrathin electronic devices have been used for flexible electronics, foldable energy devices which is used in stretchable wearable biomedical devices along with the ability to move within the body in curved geometries by which various biosignals are closely visualized. Furthermore, significant efforts were made to explore materials that experience efficient flexibility, which was achieved using graphene. The unique electromechanical, physiochemical properties of graphene have become an extreme demanding multifunctional material for flexible devices. One of the unique features of graphene is its electronic properties having high electronic mobility (15000–200000 $cm^2/(V \cdot s)$). Graphene has been used as the transparent conducting electrode TCE in different flexible applications such as touch-panels, light-emitting diodes (LEDs), photovoltaic cells, sensors, and energy harvesting devices, etc.

6.1 Touch-Based Flexible Screens

Nowadays, all smart phones are touch based. When someone touches the screen, the display takes the command and then responds on the position of the touch. Different types of touch operations include resistive, capacitive, projected capacitance, surface capacitance and surface acoustic wave types. Resistive and capacitive type operations are more frequently used on a commercial level, although such operations require the electrode materials with sheet resistance and transparency. For instance, Bae et al. prepared the graphene film having low values of sheet resistance and good transparency [112]. Due to unique properties of graphene, they were utilized for fabrication of the touch screen panel. When graphene-based touch screen was compared to ITO-based touch screen, it was observed that graphene-based touch

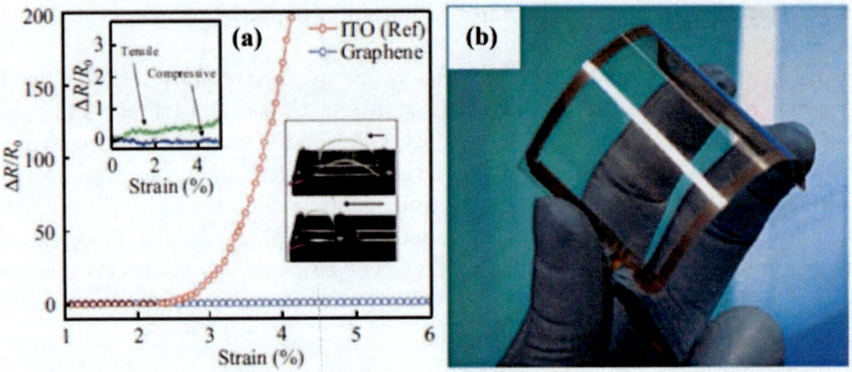

Fig. 4 Schematic diagram illustrates the **a** comparison of tensile strain of ITO and graphene and **b** flexibility and transparency of graphene respectively. Reproduced with permission from Xue et al. [132]

screen exhibited a more significant tensile strain [131] Illustrated in Fig. 4a and b the comparison of tensile strain of both and folding behavior of graphene with sharp transparency was demonstrated [132].

6.2 Flexible Organic Light Emitting Diodes

LED are those devices which transform the electrical energy into light energy of specific wavelength when a current is applied. In terms of display applications, organic based LEDs (OLEDs) have excellent features such as its display are brighter, require less power for operation, elastic, flimsy and thinner. OLEDs consist of an active layer sandwiched between the anode and cathode and among them one electrode should be transparent so the light can be visible. The active layer consists of an electroluminescent polymeric layer.

There are many studies for OLEDs where graphene was used as the transparent electrode. For instance, Wu et al. used rGO films as anode in the preparation of OLEDs. A comparison was performed using rGO-based OLEDs and ITO-based OLED devices and it was observed that the rGO-based OLED device exhibited the luminescence of 300 cd/m^2 at 11.7 V, whereas, the ITO-based OLED devices showed luminescence at 9.9 V [133]. Moreover, Chang et al. prepared the anodic material of the graphene-based composite having PEDOT as an active material [134]. The anodic material of OLEDs was also prepared by doping graphene with nitric acid (HNO_3) or gold trichloride ($AuCl_3$). A comparison of the performances was conducted, and it was observed that doped graphene films stacked in four layers were excellent as compared to ITO-based devices. Different combinations for electrodes such as for anode were developed using TiO_2/Graphene/GraHIL, graphene/ GraHIL and

Table 4 Different graphene electrodes used on organic light emitting electrodes

Devices	Medium	Electrode	Substrate	Effective current [cd A^{-1}]	References
Organic light emitting diodes (OLEDs)	Graphene sheets	+Electrode placed at Base	Glass	–	[135]
	Reduced graphene oxides	Base—electrode	Glass	7 @ 24.2 V	[136]
	CVD	+Electrode placed at Base	Polyethylene terephthalate	89.7 (max)	[137]
	CVD	+Electrode placed at Base	Glass	= 91.9 (max)	[138]
	CVD	+Electrode placed at Base	Glass	= 168.4 (max)	[139]
	CVD	+Electrode placed at Base	Glass	56.42 (max)	[140]
	CVD	+Electrode placed at Base	Plastic	>250	[140]
	CVD	+electrode placed at Base	Glass	64 @ 1000 cd m^{-2}	[141]
	CVD	+Electrode placed at Base	Glass	–	[142]
	CVD	Base—electrode	Glass	≈40 (max)	[143]
	CVD	+Electrode placed at Base	PET	82 (max)	[144]
	CVD	Base—electrode	Glass	7.9 (max)	[145]
	CVD	+electrode placed at Base	SU-8/NOA63	31.4 (max)	[146]
	CVD	+Electrode placed at Base	SiO_2/glass	–	[147]
	CVD	+Electrode placed on top	Glass	–	[148]

ITO (185 nm) /GraHIL assembly [132]. In addition, in Table 4, is the summary of different graphene electrodes used in organic light emitting diodes is illustrated.

7 Graphene-Based Materials for Biomedical Applications

Graphene is being used frequently in the last few years owning to its superior properties in a variety of different fields. As illustrated in Fig. 5, not only graphene but GO and its reduced form rGO have attained more interest in all fields.

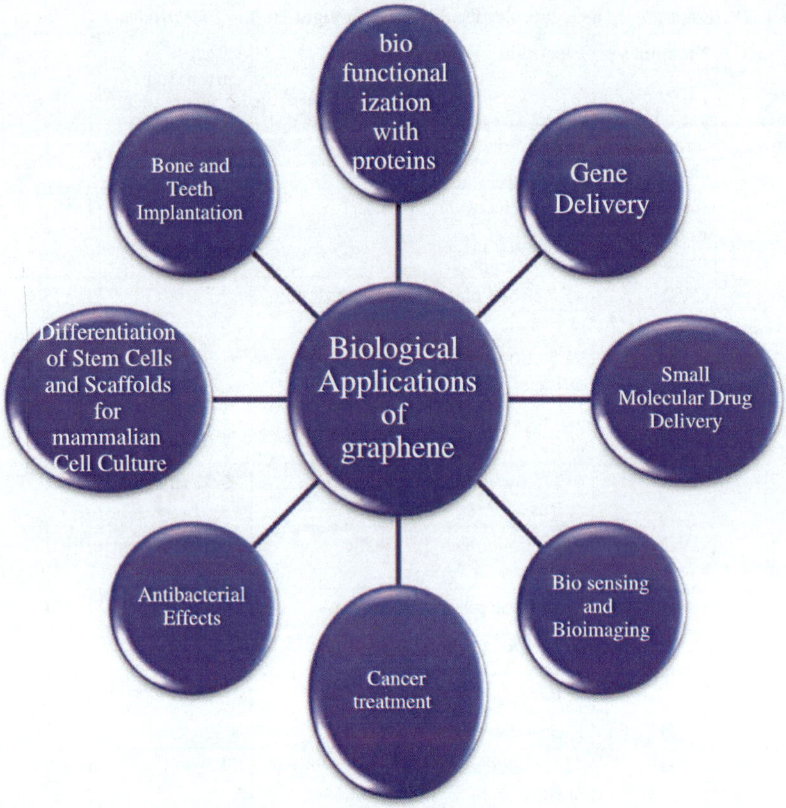

Fig. 5 Numerous applications of graphene in the biological field [149]

7.1 Biofunctionalization with Proteins

Graphene and its derivatives interact with proteins by interlinking with the secondary structure or functional group or by means of physical adsorption. The GO has oxygen in it which helps out biocatalysts and proteins to deactivate without the addition of coupling reagents [150]. Furthermore, protein absorption was monitored by using the electrolytic gate of graphene. Lots of enzymes when on activation become harmful for the body, so in certain cases, it becomes necessary to seize them. Enzymes which include lysozyme and horseradish peroxidase enzymes, can be deactivated when interacted with GO on their surface. For instance, magnetic GO has been used to deactivate laccase. By immobilization of graphene with an enzyme, the specificity of that particular enzyme increases, and a low catalytic characteristic is observed by which the enzyme can be reused. GO is also used for the deactivation of β-amylase [151]. The deactivation of different enzymes was characterized by using SPR, SEM and TEM and immobilization efficiency of 84% was observed [152]. In short, the interaction of graphene with proteins plays an important role to crystallize proteins which is helpful in studying the structure of proteins by X-ray crystallography [149].

7.2 Graphene-Based Material for DNA

Graphene-based material is used for introducing foreign DNA inside the cell. Various genetic diseases can be cured by adopting this procedure. To deliver genes, modified GOs are widely used. GO composites made of polyethylenimine (PEI) are used for cellular gene delivery via electrostatic interaction for plasmid DNA (pDNA) stacking [153]. Iron oxide nanoparticles making graphene oxide (GO) magnetic in nature has made GO a multifunctional for medicinal applications. Chitosan complexed GO (CS–GO) composite has been used for effective transportation of anticancer drugs and stacked plasmid DNA [154]. Illustrated in Fig. 6 is the significance of graphene for the drug delivery process [149]. The following Table 5 summarizes the use of graphene-based nanocomposites for DNA.

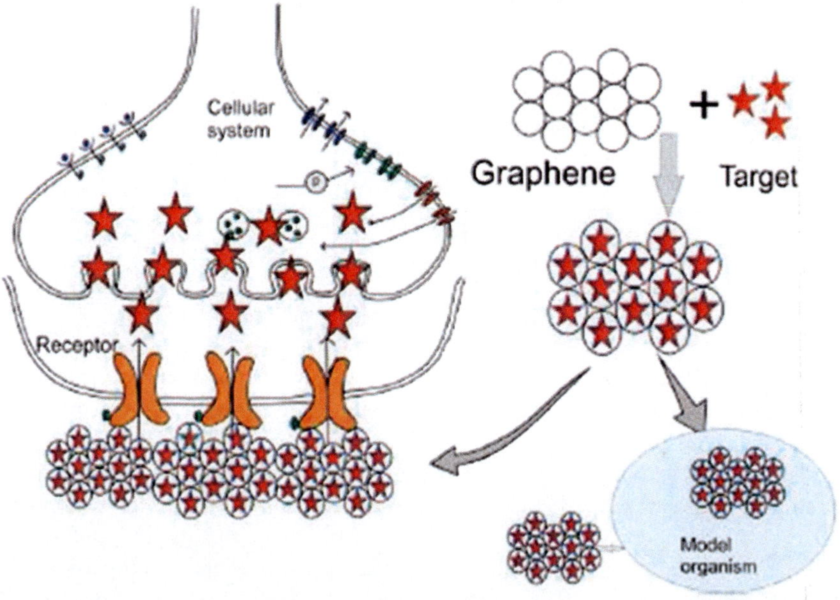

Fig. 6 Schematic model illustrates the graphene that serves as a container for target delivery. Reproduced with permission from He et al. [80]

Table 5 Graphene-based composites for delivery of genes

Graphene-based nanomaterial	Gene	Target cell in the study	References
Composite of polyethylenimine with graphene (25 kDa)	EGFP	HeLa cells	[155]
Composite of polyethylenimine with graphene e(25 kDa)	Luciferase reporter Gene	HeLa cells	[155]
Graphene oxide coated with –chitosan	Luciferase reporter Gene	HeLa cells	[156]
Composite of branched polyethylenimine with graphene oxide	Luciferase reporter Gene	HeLa and PC-3 cell lines	[157]
Grafted ultra-small Graphene oxide composite made with polyethyleneimine	EGFP	H293T and U2Os cell lines	[158]
Composite of Graphene polyethylenimine (1, 2and 10 kDa)	Enhanced green Fluorescent protein (EGFP)	HeLa cells	[159]
Polyethylenimine with GO-gold nanorods	Luciferase reporter Gene and EGFP	HeLa cells	[160]
RGO-PEG polyethylenimine	Enhanced green Fluorescent protein (EGFP)	HeLa cells	[161]

7.3 Graphene-Based Material for Biosensing and Bioimaging

Graphene and its derivatives such as GO and rGO and doped graphene (graphene composite) have been extensively used in detecting and sensing biomolecules. The biomolecules includes thrombin [153], oligonucleotides [154], ATP [151], amino corrosives [162] and dopamine [163]. A variety of GO-based biosensors have been developed. Graphene properties such as fluorescence quenching behavior of graphene and its electric property have been used for FRET-based biosensors [164]. GO-based biosensors have been developed due to outstanding physiochemical properties such as great surface area, excellent electrical conductivity and the sharp capability of gathering different biomolecules physically or chemically [149].

Bio-imaging techniques, for determining structure based on graphene, are used that includes fluorescence/confocal imaging, surface-enhanced Raman scattering (SERS), coherent anti-Stokes Raman scattering imaging (CARS), magnetic resonance imaging (MRI), positron-emission tomography (PET) [165], ultrasound imaging, photoacoustic imaging and electron paramagnetic resonance imaging (EPRI). By using all these techniques, where graphene-based materials are used, great imaging details of multiple processes involved in living cells, tissues and the whole body are provided. The most frequently used imaging technique is fluorescence imaging. A contrasting agent is used to develop the image of a body. Structures which include graphene, GO and graphene-based composites are used for bio-imaging with modification of previously used contrast agents [166].

Graphene or graphene oxide-based substrates usually are converted into carbon/G QDs of the size of 10 nm. Their fluorescence is dependent on its size and surface chemistry. The similarity of these structures can be attained by functionalizing the surface. Some of the mostly used biopolymers are polyethylene glycol (PEG), polypeptides, polyethyleneimine and polystyrene [167].

7.4 Graphene-Based Materials for Cancer Treatment

Graphene and its derivatives have been used to identify the starting level of cancer cells [168]. A tumor sphere is a solid formation developed from the budding of a cancer stem. GO inhibits the development of tumor lump efficiently in different cancer cell lines which include ovarian cancer cell lines, pancreatic cancer cell lines, breast cancer cell lines and lung cancers. Graphene and GO responses were observed in cancer cells where they induced antitumor effects and caused autophagy [169].

7.5 Graphene-Based Material for Antibacterial Effects

GO sheets and their composites made with silver nanoparticles were prepared and tested against microbes. Individually, the GO exhibited limited antimicrobial activity but the GO–Ag composite exhibited an enhanced antibacterial activity [170, 171]. For instance, Bao et al. observed the antibacterial property of GO. Here, it was revealed that the rGO-based paper/sheet reduced the growth of both Gram-negative bacteria and Gram-positive bacteria, respectively [172]. In comparison to graphene and its derivatives that include GO, Gt, rGO subordinates, it was observed that GO and rGO enhanced the antibacterial activity [173].

7.6 Graphene-Based Material for Drug Delivery

Dai et al. was the first researcher who performed the drug delivery using graphene. The purpose here was to transfer the drug at the required spot with standard biocompatibility, less virulent and distinctive targeting [172]. Moreover, Jin et al. synthesized graphene oxide hybrid functionalized with hematin-conjugated dextran for delivery of doxorubicin (DOX) [174]. Also, Zhou and his coworkers developed composites of graphene/Fe_3O_4 for loading DOX [175]. Using graphene in comparison with the

Table 6 Graphene-based materials used for delivery of drugs

Drug delivery system	Drug loaded	References
Graphene oxide-PEG- Rituxan	Doxorubicin	[176]
Graphene oxide-PEG	SN38	[177]
Graphene oxide coated with chitosan	Camptothecin	[156]
GO-Fe_3O_4-folic acid	Doxorubicin	[156]
RGO gold Nanocluster	Doxorubicin	[178]
Graphene oxide-folic acid	Doxorubicin +camptothecin	[179]
GO-PEG	Doxorubicin	[180]
GO-polyvinyl alcohol	Camptothecin	[181]
GO adamantanylporphyrin folic acid-β-cyclodextrin	Doxorubicin	[156]
GO-PEG	Doxorubicin	[182]
GO poly(Nisopropyl acrylamide)	Camptothecin	[183]
Graphene nanosheets-PF127	Doxorubicin	[181]

graphene/Fe_3O_4 composite, 65% loading capacity was obtained by graphene through π-π interaction. Graphene-based nanoparticles have been used for delivery of drugs such as Cisplatin, camptothecin, ibuprofen, 5-flurouracil, hypocrellin. Graphene and GO materials have been remarkably used as carriers for drugs for stacking a variety of antibodies, drugs used for cancer treatment, DNA, RNA, peptides, and genes. Illustrated in Table 6 are the various types of graphene-based materials that are used to deliver a drug in the body [166].

7.7 *Graphene and Graphene Oxide in Tissue Engineering*

Engler and his coworkers transfigured the tissue engineering field in 2006 by developing three different polyacrylamide matrices which were used to induce the transformed stem cells in the brain, muscle, and bone lineages. The cell differentiation is dependable upon the hardness of the substrate matrix. Graphene and its derivatives have been used in wound healing, stem cell engineering due to its unique mechanical properties such as high adaptability, strength, flexibility, and the capability to alter different functionalities on smooth surfaces. These properties of graphene made graphene a specific likely supplementary matrix in hydrogels, biodegradable films, electrospun fibers and other tissue engineering scaffolds [4] For example, when GO gets incorporated into polyvinyl acetate (PVA)-based hydrogels, its tensile strength enhances remarkably. Not only tensile strength but the compressive strength of composite hydrogels was also enhanced significantly without damaging the cytotoxicity for osteoblast cells. Graphene-based chitosan films also exhibited hardness. Along with organic polymers, biopolymers also covalently linkup with graphene-based composites [184].

At present, critical tissue damaging may lead to human casualties. Based on this fact, extensive studies have been carried out to explore graphene for stem cell engineering and musculoskeletal tissue engineering. Chen et al. studied the effect of graphene and GO for amplification and distinction of induced pluripotent stem cells (iPSCs) [185]. Here, the graphene and its derivatives encapsulated surfaces showed improved cell adhesion, cell proliferation and cell differentiation of human mesenchymal stem cells as compared to other substrates such as polydimethylsiloxane (PDMS), polyethylene terephthalate (PET), glass, and Si/SiO_2 matrix [186]. Human-adipose-derived stem cells (hASCs) can easily be detached from patients have frequently been studied in tissue engineering and Chung et al. explored the influence of GO-encapsulated substrates on hASCs [187]. It was visualized that GO film is an efficient tenet for monitoring the function of hASCs. Graphene-based materials have also been utilized for musculoskeletal tissue engineering in mouse myoblast cell lines [167].

7.7.1 Beyond 2D Substrates

2D graphene-based materials encapsulated on substrates is not sufficient for tissue engineering therefore, 3D graphene-based materials are required for the development of 3D extracellular matrix (ECM) environment. Recent progress has been done on foams of graphene used as three dimensional scaffolds for a neural stem cell (NSC) culture and human stem cell differentiation [188]. 3D graphene foams help out in fixation and feasibility of hMSCs, and urge rapid osteogenic distinction, hence making the progress of graphene-based structures for osteogenic and conductive tissue-engineered scaffolds. Shin et al. developed a 3D composite material by coating GO with gelatin methacrylate (GelMa) hydrogels. These composite scaffolds improved electromechanical properties having reduced drastic effects on coated fibroblast cells, highlighting the effectiveness of GO to be used as a nanofiller in hydrogels for three dimensional cell culture growth [189]. The reduced GO shaped into 3D porous wires were prepared for neural differentiation of neural stem cells [167].

7.8 Graphene-Based Material in Biomedical Implantation

Several different therapies and devices have been developed in the past decades for treatment of different diseases, among which metallic alloys include stainless steel and Nitinol (NiTi) have been used for implantation purpose but they lacked the cellular adhesion [190]. The extraordinary properties of graphene which include the ability to conjugate with different molecules create bio-composites with varied properties. Materials based on graphene have been used in bone and tissue engineering. Graphene and its derivatives have different properties when linked with ceramics and produce matrix with increased mechanical and osteogenic properties. Hydroxyapatite composite (Hap) is one of the materials used for bone engineering. When HAP is combined with the graphene forming HAp/GO nanocomposite incorporated with chitosan it was used as a binder with increased capabilities to release a high amount of calcium and phosphorus ions [191, 192]. Due to the availability of a large surface area of graphene, the graphene-modified bio-ceramics attained increased bioactivity. When the culture of mesenchymal cells (MSC) was uniformly scattered over a colloidal solution of HAp encapsulated with reduced graphene oxide, increased alkaline phosphates activity (ALP) and fixation of nodules of calcium over HAp were shown. Furthermore, the cells also exhibited an elevated level of expression of osteopontin and osteocalcin which are responsible for bone mineralization [193]. Graphene-based materials have been used frequently in dental implantation. When artificial acrylic teeth were coated with the nanocomposite of zinc oxide and graphene, the deposition of bacteria reduced swiftly [167].

7.9 Graphene-Based Material for Photothermal Therapy (PTT)

Photo therapy is the killing of cancer cells by using specific light radiations. The advantage of such therapies includes targeting the tumor cells without any side effects.

In PTT, photosensitizing factors are introduced in the body to go to the required cancer cells where the light is absorbed leading photoablation of cancer cells which kills these cells. To investigate such procedures, a variety of different nanomaterials have been explored such as nano-shells, nano-rods, single-walled carbon nanotubes (SWCNTs), multi-walled carbon nanotubes (MWCNTs) and gold pyramids. Robinson and his coworker used GO composite made of rGO PEGylated and selectively performed photo ablation of cancer cells at a small scale where graphene was incorporated with gold nanocrystals and is used for (Doxycycline) DOX delivery and by combining PT with imaging against breast cancer cells [166].

8 Graphene Based Materials for Water Purification

Natural water resources such as rivers, lakes, groundwater, and oceans are getting polluted by different contaminants such as heavy as well as radioactive metals, organic pollutants including dyes as well pigments, agicultural pollutants including pesticides, herbicides and optiods. These contaminants are generated from the wastewater discharged from industries, agriculture, and sewage which affect our health and environment adversely. Different technologies and advancements are required for the purification of water required for a healthy community. The limit of contaminants designed by World health organization (WHO) and environment protection agency EPA has exceeded in water. Heavy metals, metal containing compounds and radionuclides present in water may enter the human body when potable water is consumed. For purification of water one of the methods used is the adsorption. Among various adsorbents Graphene and GO have been proved adsorbents due to their unique physiochemical characteristics such as elevated surface area, conductivity and elevated sorption capability that are broadly used in water purification [194].

8.1 Heavy Metals Removal

Among water pollutants, heavy metals are the vital contaminants. Consumption of these heavy metals has adverse health risks to human. Many researchers used magnetic materials for water pollution treatment. For instance, Li and his coworkers and Xu et al. used derivatives of magnetic inverse spinel ferrites for purification of water due to characteristics such as magnetic and chemical sustainability, and elevated surface porous structure [195]. The nano-metal ferrites have poor stability due to which graphene-based composites having magnetic ferrites as an active material was prepared [194].

The magnetic organic composites of GO were used for the separation of precious metals. The organic functional groups such as carboxylic group, hydroxyl group and epoxy group improved the adsorption of heavy metals. For water treatment, composites of magnetic GO (MGO) have shown high performance. Additionally, Chandra et al. prepared superparamagnetic rMGO composites which resulted in high sorption capability above 99.9% for quenching As(III) and As(V) [196]. Furthermore, Luo et al. prepared Fe_3O_4-MnO-based rGO composites for the efficient sorption of As(III) and As(V) [197]. Zhang et al. developed a graphene composite based on ferric hydroxide for maximum sorption of arsenate at the pH range of between 4 and 9 from contaminated water, where the concentration of arsenate reduced to 20–0.5 ppm [194].

MGO nanocomposites showed high efficiency in sorption of Cd(II), Pb(II), and Cu(II) due to their unique features such as high surface area, sharp complexation ability, and excellent quenching ability [198]. Composites of RGO–MnO_2 showed excellent mercury quenching capability. For instance, Liu et al. successfully pulled out Co(II) using MGO. Moreover, the same researcher developed a MGO composite having 3-aminopropyltrimethoxysilane as the active material that showed improved adsorption capability for Cr(VI) [194].

Bhunia et al. created a heterogeneous matrix, for the effective quenching of heavy metals, made up of the graphene iron/iron oxide composite (rGO-FeO/Fe_3O_4) [199]. For efficient quenching of Cr(VI), Wang et al. developed a graphene composite where polypyrrole was encapsulated with rMGO [200].

8.2 Radioactive Metal Ions Removal

The activities such as production of nuclear power products and mining have delivered nuclear discharge into the aquatic domain. These radioactive materials are responsible for long-term threat to the environment. The explosion in Japan at the Fukushima Daiichi Nuclear plant delivered artificial radionuclides which caused disability in humans. Zhao et al. prepared graphene nanosheets which exhibited a highest sorption capability (qmax) of 97.5 mg/g for U(VI) ions. But due to the assortment of nanosheets in water where a decrease in efficiency was observed. This was overcome by using the magnetically functionalized GO nanosheets [194].

Sun et al. prepared the nanosized iron-rGO based composite for quenching of U(VI) [201]. The higher absorption value of U(VI) was interlinked with the acidity of the solution where positively charged species of U(VI) were produced while in basic media, and negatively charged species reduced its adsorption efficiency. Zong et al. developed the composite of graphene oxide with Fe_3O_4 for the elimination of U(VI) at qmax of 69.49 mg/g [202]. Zhao et al. prepared the graphene composite coated with amidoximated magnetite/GO (AOMGO) and visualized the adsorption of U(VI) resulted for a high qmax of 284.92 mg/g [203]. The composite of rGO made up with MnO_2–Fe_3O_4 was successfully fabricated by Tan and his coworkers and employed for the abolition of U(VI) via adsorption which resulted in 108.7 mg/g U(VI) abolition and this composite loaded with radionucleotide could be successfully reprocessed and reutilized [204]. Alternatively, Lingamdinne and his coworkers prepared magnetic graphene nanocomposite of nickel ferrite-GO for the successful removal of U(VI), and Th(IV). The studies explored that these magnetic nanocomposites could be regenerated and processed successfully up to five cycles [194].

8.3 Organic Pollutants Removal

8.3.1 Dyes and Pigments

Dyes and pigments are frequently used in the products of different industries such as paper and pulp, fabric, varnishes, plastic, and leather industries. These industries dispose off excess of dyes and pigment as discharge in water ending up as factory discharge. These pigments (organic in nature) adversely affect the penetration of sunlight which in turn affects the photosynthesis process of aquatic plants. The pigments are also toxic materials to human and aquatic creatures as they do not undergo a chemical reaction or biodegradation, hence, being responsible for causing tumor, mutagenesis etc. and other acute problems in ecosystem. Therefore, its necessary to remove these dyes effectively before exposure to the aquatic community. Common dyes that enter in wastewater include methylene blue, methylene orange, malachite green, Congo red, BR-12, Azo acid red 14, acid orange 8, acid yellow

99, amido black [204], Persian orange, reactive orange 12, and a few more [205]. A variety of different methodologies have been utilized for successful removal of dyes such as photocatalytic degradation, separation by membrane, coagulation process, electrolysis, liquid-liquid extraction and adsorption [194].

In the past years, nanoparticles having magnetic properties have been utilized for the sorption of organic pollutants. Nanoparticles having unique features such as reduced toxicity, sharper chemical sustainability, superior magnetic properties, and extraordinary recycling capability proved to be efficient in pollutant removal. The magnetic nanoparticles exposed to air get oxidized rapidly which affects their efficiency. In order to prevent oxidation of these magnetic nanoparticles, they were encapsulated with graphene which caused successful abolition of contaminants [206]. Deng et al. synthesized MGO and utilized them against the abolition of Cd(II) and ionic pigments including orange Gelb (OG) and methylene blue (MB) [194].

In addition, Yang et al. synthesized the composite of rGO with nickel and its intermixed metal oxide (rGO/Ni/MMO) [207]. This hybrid material exhibited a sharp proficiency for abolition of methylene orange (MO) from aqueous media. Abdi et al. prepared magnetic graphene-based composite (MMGO) encapsulated with polyethersulfone (PES) polymer, which were studied for the removal of copper and dyes [208]. Khurana et al. studied the sorption of Eriochrome Black T(EBT) from textile discharge using MGO [209]. Subsequently, they also prepared MGO coated with Fe_2O_3 for the extraction of toxic azodye EBT used in the textile industry [194].

8.4 *Other Agricultural Pollutants*

8.4.1 Pesticides

The discharge of water from farms, where crops are cultivated such as maize, rice, vegetables and fruit pollute the marine community. The major contaminants include insecticides or pesticides. The most commonly used insecticide is the Neonicotinoid-based insecticide. Due to repeated usage of these insecticides, it gets accumulated by the passage of time hence these insecticides pollute soil, crops, vegetables etc. disturbing health of ecosystem. For instance, Liu et al. prepared a copper-based magnetic metal-organic framework (MMOF), encapsulated with Fe_4O_3–GO-β- cyclodextrin (β-CD) nanocomposite which was used for the extraction of insecticide, an impurity from aqueous solution [194].

8.4.2 Herbicides

In order to control weed species, selective herbicides are frequently used in farms. Liu and his coworkers prepared a magnetic polymer known as magnetic molecularly imprinted polymer (MMIP) by using g-C_3N_4–Fe_3O_4 for the absorption of atrazine from the solution [210]. Here, the results revealed that the MMIP resulted in a higher qmax and exhibited strong harmony for abolition of atrazine. Alternatively, Boruah and his coworkers developed Fe_3O_4 supported rGO nanocomposite for abolition of toxic pesticides from aquatic domain [211]. The pesticides such assimazine, simeton, atrazine prometryn, and ametryn were found to be present in aquatic domain. The research studies discovered that the Fe_3O_4/rGO nanocomposite is reusable and showed an efficient adsorption activity towards the adsorption of the pesticides molecules instead of using Fe_3O_4 nanoparticles and rGO sheets individually [194].

8.4.3 Opioids

One of the opioids, Methadone, its market name is dolophine, is used for the treatment of pain, as a maintenance therapy. Gupta et al. reported the adsorption capacity of methadone by using MGO as an adsorbent. The qmax value was observed at 87.2 mg/g for abolition of methadone [194].

9 Conclusion

The graphene, being a novel material, has been exposed to a variety of new research fields and has opened up a new horizon. Graphene is the lightest material possessing unique features such as extensive surface area, great strength, sharp electrical conductivity, excellent thermal conductivity and is easy to get functionalized, gas impenetrable and chemical stability; as a result, graphene and its derivatives are utilized in almost all fields. Graphene, GO and its reduced form, hierarchical porous 3D graphene and graphene composites have been widely used in different fields such as in optoelectronics, sensing, catalysis, energy engineering including electric energy storage, electronics and optoelectronics, in electrode and asymmetric capacitor electrode material, solar-thermal energy, in the field of medicine including delivery of drug, therapies for gene, cancer, photothermal and photodynamic, tissue engineering, bioimaging, nanomedicine, and biomolecular sensing. In this chapter, the detailed applications of the graphene family have been covered and discussed. For obtaining high performance of graphene in different fields, the development of graphene functionality is in progress for future prospective.

References

1. Kumar, S.S.A., et al.: New perspectives on Graphene/Graphene oxide based polymer nanocomposites for corrosion applications: the relevance of the Graphene/Polymer barrier coatings. Prog. Org. Coat. **154**, 106215 (2021)
2. Wan, X., Huang, Y., Chen, Y.: Focusing on energy and optoelectronic applications: a journey for graphene and graphene oxide at large scale. Acc. Chem. Res. **45**(4), 598–607 (2012)
3. Kumar, S.S.A., et al.: A comprehensive review: super hydrophobic graphene nanocomposite coatings for underwater and wet applications to enhance corrosion resistance. FlatChem, 100326 (2021)
4. Sharma, S.S.A., et al.: The significance of graphene based composite hydrogels as smart materials: a review on the fabrication, properties, and its applications. FlatChem, 100352 (2022)
5. Armand, M., Tarascon, J.-M.: Building better batteries. Nature **451**(7179), 652–657 (2008)
6. Dubal, D.P., et al.: Hybrid energy storage: the merging of battery and supercapacitor chemistries. Chem. Soc. Rev. **44**(7), 1777–1790 (2015)
7. Meng, J., et al.: Advances in structure and property optimizations of battery electrode materials. Joule **1**(3), 522–547 (2017)
8. Geim, A.K., Novoselov, K.S.: The rise of graphene. In: Nanoscience and technology: a collection of reviews from nature journals, pp. 11–19. World Scientific (2010)
9. Parviz, D., et al.: Challenges in liquid-phase exfoliation, processing, and assembly of pristine graphene. Adv. Mater. **28**(40), 8796–8818 (2016)
10. Sygletou, M., et al.: Advanced photonic processes for photovoltaic and energy storage systems. Adv. Mater. **29**(39), 1700335 (2017)
11. Xiang, Z., et al.: Edge functionalization of graphene and two-dimensional covalent organic polymers for energy conversion and storage. Adv. Mater. **28**(29), 6253–6261 (2016)
12. Liu, J.: Charging graphene for energy. Nat. Nanotechnol. **9**(10), 739–741 (2014)
13. Bonaccorso, F., et al.: Graphene, related two-dimensional crystals, and hybrid systems for energy conversion and storage. Science **347**(6217), 1246501 (2015)
14. Raccichini, R., et al.: The role of graphene for electrochemical energy storage. Nat. Mater. **14**(3), 271–279 (2015)
15. El-Kady, M.F., Shao, Y., Kaner, R.B.: Graphene for batteries, supercapacitors and beyond. Nat. Rev. Mater. **1**(7), 1–14 (2016)
16. Renteria, J.D., Nika, D.L., Balandin, A.A.: Graphene thermal properties: applications in thermal management and energy storage. Appl. Sci. **4**(4), 525–547 (2014)
17. Kiessling, D., et al.: Novel nanographene/porphyrin hybrids–preparation, characterization, and application in solar energy conversion schemes. Chem. Sci. **4**(8), 3085–3098 (2013)
18. Hou, J., et al.: Graphene-based electrochemical energy conversion and storage: fuel cells, supercapacitors and lithium ion batteries. Phys. Chem. Chem. Phys. **13**(34), 15384–15402 (2011)
19. Choi, D., et al.: Fully rollable transparent nanogenerators based on graphene electrodes. Adv. Mater. **22**(19), 2187–2192 (2010)
20. Shin, H.-J., et al.: Control of electronic structure of graphene by various dopants and their effects on a nanogenerator. J. Am. Chem. Soc. **132**(44), 15603–15609 (2010)
21. Choi, H.-J., et al.: Graphene for energy conversion and storage in fuel cells and supercapacitors. Nano Energy **1**(4), 534–551 (2012)
22. Lv, X.-J., et al.: Hydrogen evolution from water using semiconductor nanoparticle/graphene composite photocatalysts without noble metals. J. Mater. Chem. **22**(4), 1539–1546 (2012)
23. Tu, X., et al.: One-pot synthesis, characterization, and enhanced photocatalytic activity of a BiOBr–graphene composite. Chem.–A Eur. J. **18**(45), 14359–14366 (2012)
24. Cao, X., et al.: Three-dimensional graphene network composites for detection of hydrogen peroxide. Small **9**(9–10), 1703–1707 (2013)
25. Wei, W., et al.: Nanocomposites of graphene oxide and upconversion rare-earth nanocrystals with superior optical limiting performance. Small **8**(14), 2271–2276 (2012)

26. Wu, Q., et al.: Supercapacitors based on flexible graphene/polyaniline nanofiber composite films. ACS Nano **4**(4), 1963–1970 (2010)
27. Lu, X., et al.: Polypyrrole/carbon nanotube nanocomposite enhanced the electrochemical capacitance of flexible graphene film for supercapacitors. J. Power Sources **197**, 319–324 (2012)
28. Pei, S., et al.: Direct reduction of graphene oxide films into highly conductive and flexible graphene films by hydrohalic acids. Carbon **48**(15), 4466–4474 (2010)
29. Wang, Z., et al.: Recent advances in 3D graphene architectures and their composites for energy storage applications. Small **15**(3), 1803858 (2019)
30. Liu, F., Zhu, J., Xue, D.: MnO_2-graphene nanocomposites by ripening of amorphous MnO_2 in mild conditions. Graphene **1**(1), 58–62 (2013)
31. Liu, F., Xue, D.: Advanced graphene nanomaterials for electrochemical energy storage. Mater. Res. Innov. **19**(1), 7–19 (2015)
32. Liu, F., Zhu, J., Xue, D.: Fabrication of MnO_2-graphene nanocomposite by ripening of amorphous MnO_2 in graphene oxide matrix. Sci. Adv. Mater. **5**(7), 904–908 (2013)
33. Chen, K., Xue, D.: In-situ electrochemical route to aerogel electrode materials of graphene and hexagonal CeO_2. J. Colloid Interface Sci. **446**, 77–83 (2015)
34. Hooch Antink, W., et al.: Recent progress in porous graphene and reduced graphene oxide-based nanomaterials for electrochemical energy storage devices. Adv. Mater. Interfaces **5**(5), 1701212 (2018)
35. Chang, H., Wu, H.: Graphene-based nanocomposites: preparation, functionalization, and energy and environmental applications. Energy Environ. Sci. **6**(12), 3483–3507 (2013)
36. Zhao, C., et al.: Recent development of fabricating flexible micro-supercapacitors for wearable devices. Adv. Mater. Technol. **3**(9), 1800028 (2018)
37. Kou, Z., et al.: Molybdenum carbide-derived chlorine-doped ordered mesoporous carbon with few-layered graphene walls for energy storage applications. ACS Appl. Mater. Interfaces **9**(4), 3702–3712 (2017)
38. Zhao, Y., et al.: NPO co-doped high performance 3D graphene prepared through red phosphorous-assisted "cutting-thin" technique: a universal synthesis and multifunctional applications. Nano Energy **28**, 346–355 (2016)
39. Wang, B., et al.: Graphene-based composites for electrochemical energy storage. Energy Storage Mater. **24**, 22–51 (2020)
40. Mo, R., et al.: 3D nitrogen-doped graphene foam with encapsulated germanium/nitrogen-doped graphene yolk-shell nanoarchitecture for high-performance flexible Li-ion battery. Nat. Commun. **8**(1), 1–9 (2017)
41. Sun, H., et al.: Three-dimensional holey-graphene/niobia composite architectures for ultrahigh-rate energy storage. Science **356**(6338), 599–604 (2017)
42. Fan, X., Chen, X., Dai, L.: 3D graphene based materials for energy storage. Curr. Opin. Colloid Interface Sci. **20**(5–6), 429–438 (2015)
43. Gwon, H., et al.: Flexible energy storage devices based on graphene paper. Energy Environ. Sci. **4**(4), 1277–1283 (2011)
44. Lv, W., et al.: Graphene-based materials for electrochemical energy storage devices: opportunities and challenges. Energy Storage Mater. **2**, 107–138 (2016)
45. Bruce, P.G., et al.: Li–O_2 and Li–S batteries with high energy storage. Nat. Mater. **11**(1), 19–29 (2012)
46. Yu, M., et al.: Graphene materials for lithium–sulfur batteries. Energy Storage Mater. **1**, 51–73 (2015)
47. Xu, J., et al.: Sulfur–graphene nanostructured cathodes via ball-milling for high-performance lithium–sulfur batteries. ACS Nano **8**(10), 10920–10930 (2014)
48. Hwa, Y., Zhao, J., Cairns, E.J.: Lithium sulfide (Li_2S)/graphene oxide nanospheres with conformal carbon coating as a high-rate, long-life cathode for Li/S cells. Nano Lett. **15**(5), 3479–3486 (2015)
49. Gu, X., et al.: A porous nitrogen and phosphorous dual doped graphene blocking layer for high performance Li–S batteries. J. Mater. Chem. A **3**(32), 16670–16678 (2015)

50. Zhou, W., et al.: Amylopectin wrapped graphene oxide/sulfur for improved cyclability of lithium–sulfur battery. ACS Nano **7**(10), 8801–8808 (2013)
51. Xu, C., et al.: Sulfur/three-dimensional graphene composite for high performance lithium–sulfur batteries. J. Power Sources **275**, 22–25 (2015)
52. Wang, J.-G., Xie, K., Wei, B.: Advanced engineering of nanostructured carbons for lithium–sulfur batteries. Nano Energy **15**, 413–444 (2015)
53. Tang, C., et al.: CaO-templated growth of hierarchical porous graphene for high-power lithium–sulfur battery applications. Adv. Func. Mater. **26**(4), 577–585 (2016)
54. Liu, M., et al.: Nano Res. **9**, 3735–3746 (2016); (g) M. Liu, B. Li, H. Zhou, C. Chen, Y. Liu, T. Liu, Chem. Commun. **53**, 2810–2813 (2017)
55. Huang, J.-Q., et al.: Three-dimensional porous graphene aerogel cathode with high sulfur loading and embedded TiO_2 nanoparticles for advanced lithium–sulfur batteries. ACS Appl. Mater. Interfaces **8**(42), 28663–28670 (2016)
56. Tang, C., et al.: Nitrogen-doped aligned carbon nanotube/graphene sandwiches: facile catalytic growth on bifunctional natural catalysts and their applications as scaffolds for high-rate lithium-sulfur batteries. Adv. Mater. **26**(35), 6100–6105 (2014)
57. Zhou, G., et al.: High-performance lithium-sulfur batteries with a self-supported, 3D Li_2S-doped graphene aerogel cathodes. Adv. Energy Mater. **6**(2), 1501355 (2016)
58. Luo, X.-F., Yang, C.-H., Chang, J.-K.: Correlations between electrochemical Na+ storage properties and physiochemical characteristics of holey graphene nanosheets. J. Mater. Chem. A **3**(33), 17282–17289 (2015)
59. Rui, X., et al.: An advanced sodium-ion battery composed of carbon coated Na_3V_2 $(PO_4)_3$ in a porous graphene network. Adv. Mater. **27**(42), 6670–6676 (2015)
60. Xu, J., et al.: Nitrogen-doped holey graphene as an anode for lithium-ion batteries with high volumetric energy density and long cycle life. Small **11**(46), 6179–6185 (2015)
61. Yang, D., et al.: Three-dimensional nitrogen-doped holey graphene and transition metal oxide composites for sodium-ion batteries. J. Mater. Chem. A **7**(1), 363–371 (2019)
62. Wang, Y.-X., et al.: Ultrafine SnO_2 nanoparticle loading onto reduced graphene oxide as anodes for sodium-ion batteries with superior rate and cycling performances. J. Mater. Chem. A **2**(2), 529–534 (2014)
63. Liu, H., et al.: 3D–0D graphene-Fe_3O_4 quantum dot hybrids as high-performance anode materials for sodium-ion batteries. ACS Appl. Mater. Interfaces **8**(40), 26878–26885 (2016)
64. Zhang, X., et al.: $MgFe_2O_4$/reduced graphene oxide composites as high-performance anode materials for sodium ion batteries. Electrochim. Acta **180**, 616–621 (2015)
65. Qu, B., et al.: Layered SnS_2-reduced graphene oxide composite–a high-capacity, high-rate, and long-cycle life sodium-ion battery anode material. Adv. Mater. **26**(23), 3854–3859 (2014)
66. Liu, Y., et al.: Liquid phase exfoliated MoS2 nanosheets percolated with carbon nanotubes for high volumetric/areal capacity sodium-ion batteries. ACS Nano **10**(9), 8821–8828 (2016)
67. Wan, F., et al.: In situ binding Sb nanospheres on graphene via oxygen bonds as superior anode for ultrafast sodium-ion batteries. ACS Appl. Mater. Interfaces **8**(12), 7790–7799 (2016)
68. Song, J., et al.: Chemically bonded phosphorus/graphene hybrid as a high performance anode for sodium-ion batteries. Nano Lett. **14**(11), 6329–6335 (2014)
69. Sun, J., et al.: A phosphorene–graphene hybrid material as a high-capacity anode for sodium-ion batteries. Nat. Nanotechnol. **10**(11), 980–985 (2015)
70. Fang, Y., et al.: 3D graphene decorated $NaTi_2$ $(PO_4)_3$ microspheres as a superior high-rate and ultracycle-stable anode material for sodium ion batteries. Adv. Energy Mater. **6**(19), 1502197 (2016)
71. Zhu, H., et al.: Free-standing $Na_{2/3}Fe_{1/2}Mn_{1/2}O_2$@ Graphene film for a sodium-ion battery cathode. ACS Appl. Mater. Interfaces **6**(6), 4242–4247 (2014)
72. Olabi, A.G., et al.: Application of graphene in energy storage device–a review. Renew. Sustain. Energy Rev. **135**, 110026 (2021)
73. Tao, Y., Sui, Z.-Y., Han, B.-H.: Advanced porous graphene materials: from in-plane pore generation to energy storage applications. J. Mater. Chem. A **8**(13), 6125–6143 (2020)

74. Wu, C., Maier, J., Yu, Y.: Sn-based nanoparticles encapsulated in a porous 3D graphene network: advanced anodes for high-rate and long life Li-ion batteries. Adv. Func. Mater. **25**(23), 3488–3496 (2015)
75. Jiang, L., et al.: Monolayer MoS2–graphene hybrid aerogels with controllable porosity for lithium-ion batteries with high reversible capacity. ACS Appl. Mater. Interfaces **8**(4), 2680–2687 (2016)
76. Yu, H., et al.: Designed synthesis of ordered mesoporous graphene spheres from colloidal nanocrystals and their application as a platform for high-performance lithium-ion battery composite electrodes. Nano Res. **9**(12), 3757–3771 (2016)
77. Lee, S.-H., et al.: Nanohole-structured, iron oxide-decorated and gelatin-functionalized graphene for high rate and high capacity Li-Ion anode. Carbon **119**, 355–364 (2017)
78. Shi, J.L., et al.: 3D mesoporous graphene: CVD self-assembly on porous oxide templates and applications in high-stable Li-S batteries. Small **11**(39), 5243–5252 (2015)
79. Hu, G., et al.: 3D graphene-foam–reduced-graphene-oxide hybrid nested hierarchical networks for high-performance Li–S batteries. Adv. Mater. **28**(8), 1603–1609 (2016)
80. He, J., et al.: Three-dimensional CNT/graphene–sulfur hybrid sponges with high sulfur loading as superior-capacity cathodes for lithium–sulfur batteries. J. Mater. Chem. A **3**(36), 18605–18610 (2015)
81. Lu, Y., et al.: High-capacity and ultrafast Na-ion storage of a self-supported 3D porous antimony persulfide–graphene foam architecture. Nano Lett. **17**(6), 3668–3674 (2017)
82. Wang, Z.L., et al.: Graphene oxide gel-derived, free-standing, hierarchically porous carbon for high-capacity and high-rate rechargeable Li-O_2 batteries. Adv. Func. Mater. **22**(17), 3699–3705 (2012)
83. Wang, Y., et al.: Preventing graphene sheets from restacking for high-capacitance performance. J. Phys. Chem. C **115**(46), 23192–23197 (2011)
84. Ma, J., et al.: Synthesis of C60/graphene composite as electrode in supercapacitors. Fuller. Nanotub. Carbon Nanostruct. **23**(6), 477–482 (2015)
85. Yan, J., et al.: Electrochemical properties of graphene nanosheet/carbon black composites as electrodes for supercapacitors. Carbon **48**(6), 1731–1737 (2010)
86. Kinloch, I.A., et al.: Composites with carbon nanotubes and graphene: an outlook. Science **362**(6414), 547–553 (2018)
87. Choi, B.G., et al.: 3D macroporous graphene frameworks for supercapacitors with high energy and power densities. ACS Nano **6**(5), 4020–4028 (2012)
88. Saikia, B.K., et al.: A brief review on supercapacitor energy storage devices and utilization of natural carbon resources as their electrode materials. Fuel **282**, 118796 (2020)
89. Xu, Y., et al.: Solution processable holey graphene oxide and its derived macrostructures for high-performance supercapacitors. Nano Lett. **15**(7), 4605–4610 (2015)
90. Bai, Y., et al.: Formation process of holey graphene and its assembled binder-free film electrode with high volumetric capacitance. Electrochim. Acta **187**, 543–551 (2016)
91. Kim, H.-K., et al.: Scalable fabrication of micron-scale graphene nanomeshes for high-performance supercapacitor applications. Energy Environ. Sci. **9**(4), 1270–1281 (2016)
92. Kannan, A.G., Samuthirapandian, A., Kim, D.-W.: Electric double layer capacitors employing nitrogen and sulfur co-doped, hierarchically porous graphene electrodes with synergistically enhanced performance. J. Power Sources **337**, 65–72 (2017)
93. Wang, Y., et al.: Generation of three dimensional pore-controlled nitrogen-doped graphene hydrogels for high-performance supercapacitors by employing formamide as the modulator. J. Mater. Chem. A **5**(4), 1442–1445 (2017)
94. Xu, J., et al.: A hierarchical carbon derived from sponge-templated activation of graphene oxide for high-performance supercapacitor electrodes. Adv. Mater. **28**(26), 5222–5228 (2016)
95. Zhang, N., et al.: Three-dimensional pompon-like MnO_2/graphene hydrogel composite for supercapacitor. Electrochim. Acta **210**, 804–811 (2016)
96. Wei, B., et al.: In situ growth of manganese oxide on 3D graphene by a reverse microemulsion method for supercapacitors. J. Power Sources **307**, 129–137 (2016)

97. Pan, Z., et al.: High electroactive material loading on a carbon nanotube@ 3D graphene aerogel for high-performance flexible all-solid-state asymmetric supercapacitors. Adv. Func. Mater. **27**(27), 1701122 (2017)
98. Gao, S., et al.: Extraordinarily high-rate capability of polyaniline nanorod arrays on graphene nanomesh. J. Power Sources **304**, 111–118 (2016)
99. He, Y., et al.: Holey graphene/polypyrrole nanoparticle hybrid aerogels with three-dimensional hierarchical porous structure for high performance supercapacitor. J. Power Sources **317**, 10–18 (2016)
100. Cai, X., et al.: Solvothermal synthesis of NiCo-layered double hydroxide nanosheets decorated on RGO sheets for high performance supercapacitor. Chem. Eng. J. **268**, 251–259 (2015)
101. Zhu, Y., et al.: Carbon-based supercapacitors produced by activation of graphene. Science **332**(6037), 1537–1541 (2011)
102. Wang, H., et al.: Design of graphene-coated hollow mesoporous carbon spheres as high performance electrodes for capacitive deionization. J. Mater. Chem. A **2**(13), 4739–4750 (2014)
103. Li, H., et al.: Reduced graphene oxide and activated carbon composites for capacitive deionization. J. Mater. Chem. **22**(31), 15556–15561 (2012)
104. Yu, D., Dai, L.: Self-assembled graphene/carbon nanotube hybrid films for supercapacitors. J. Phys. Chem. Lett. **1**(2), 467–470 (2010)
105. Zhang, D., et al.: Enhanced capacitive deionization performance of graphene/carbon nanotube composites. J. Mater. Chem. **22**(29), 14696–14704 (2012)
106. Boota, M., et al.: Pseudocapacitance and excellent cyclability of 2, 5-dimethoxy-1, 4-benzoquinone on graphene. Energy Environ. Sci. **9**(8), 2586–2594 (2016)
107. Yan, J., et al.: Flexible MXene/graphene films for ultrafast supercapacitors with outstanding volumetric capacitance. Adv. Func. Mater. **27**(30), 1701264 (2017)
108. Yan, L., et al.: Single layer graphene electrodes for quantum dot-light emitting diodes. Nanotechnology **26**(13), 135201 (2015)
109. Yin, D., et al.: Mechanically robust stretchable organic optoelectronic devices built using a simple and universal stencil-pattern transferring technology. Light. Sci. Appl. **7**(1), 1–8 (2018)
110. Bi, Y.-G., et al.: Ultrathin and ultrasmooth Au films as transparent electrodes in ITO-free organic light-emitting devices. Nanoscale **8**(19), 10010–10015 (2016)
111. Zeng, B., et al.: Hybrid graphene metasurfaces for high-speed mid-infrared light modulation and single-pixel imaging. Light. Sci. Appl. 7(1), 1–8 (2018)
112. Bae, S., et al.: Roll-to-roll production of 30-inch graphene films for transparent electrodes. Nat. Nanotechnol. **5**(8), 574–578 (2010)
113. Yin, Z., et al.: Organic photovoltaic devices using highly flexible reduced graphene oxide films as transparent electrodes. ACS Nano **4**, 5263–5268 (2010)
114. Lee, S., et al.: Flexible organic solar cells composed of P3HT: PCBM using chemically doped graphene electrodes. Nanotechnology **23**(34), 344013 (2012)
115. Lee, Y.M., et al.: Incorporation of a metal oxide interlayer using a virus-templated assembly for synthesis of graphene-electrode-based organic photovoltaics. ChemSusChem **8**(14), 2385–2391 (2015)
116. Du, J., et al.: Efficient organic photovoltaic cells on a single layer graphene transparent conductive electrode using MoO x as an interfacial layer. Nanoscale **9**(1), 251–257 (2017)
117. Chen, Y., et al.: A two-step thermal annealing and HNO_3 doping treatment for graphene electrode and its application in small-molecule organic solar cells. Org. Electron. **38**, 35–41 (2016)
118. Kim, H., et al.: Organic solar cells using CVD-grown graphene electrodes. Nanotechnology **25**(1), 014012 (2013)
119. Kim, K., et al.: Ultrathin organic solar cells with graphene doped by ferroelectric polarization. ACS Appl. Mater. Interfaces **6**(5), 3299–3304 (2014)
120. La Notte, L., et al.: Sprayed organic photovoltaic cells and mini-modules based on chemical vapor deposited graphene as transparent conductive electrode. Carbon **129**, 878–883 (2018)

121. Gomez De Arco, L., et al.: Continuous, highly flexible, and transparent graphene films by chemical vapor deposition for organic photovoltaics. ACS Nano **4**(5), 2865–2873 (2010)
122. Kim, H., et al.: On-fabrication solid-state N-doping of graphene by an electron-transporting metal oxide layer for efficient inverted organic solar cells. Adv. Energy Mater. **6**(12), 1600172 (2016)
123. Kavan, L., Yum, J.H., Grätzel, M.: Optically transparent cathode for dye-sensitized solar cells based on graphene nanoplatelets. ACS Nano **5**(1), 165–172 (2011)
124. Xue, Y., et al.: Nitrogen-doped graphene foams as metal-free counter electrodes in high-performance dye-sensitized solar cells. Angew. Chem. Int. Ed. **51**(48), 12124–12127 (2012)
125. Wang, X., Zhi, L., Müllen, K.: Transparent, conductive graphene electrodes for dye-sensitized solar cells. Nano Lett. **8**(1), 323–327 (2008)
126. Roy-Mayhew, J.D., et al.: Functionalized graphene as a catalytic counter electrode in dye-sensitized solar cells. ACS Nano **4**(10), 6203–6211 (2010)
127. Jung, S., et al.: Development of annealing-free, solution-processable inverted organic solar cells with N-doped graphene electrodes using zinc oxide nanoparticles. Nano Lett. **18**(2), 1337–1343 (2018)
128. Yoon, J., et al.: Superflexible, high-efficiency perovskite solar cells utilizing graphene electrodes: towards future foldable power sources. Energy Environ. Sci. **10**(1), 337–345 (2017)
129. Chen, Y., et al.: Graphene as a transparent and conductive electrode for organic optoelectronic devices. Adv. Electron. Mater. **5**(10), 1900247 (2019)
130. Rana, K., Singh, J., Ahn, J.-H.: A graphene-based transparent electrode for use in flexible optoelectronic devices. J. Mater. Chem. C **2**(15), 2646–2656 (2014)
131. Ryu, J., et al.: Fast synthesis of high-performance graphene films by hydrogen-free rapid thermal chemical vapor deposition. ACS Nano **8**(1), 950–956 (2014)
132. Xue, Z., et al.: Engineering in-plane silicon nanowire springs for highly stretchable electronics. J. Semicond. **39**(1), 011001 (2018)
133. Wu, J., et al.: Organic light-emitting diodes on solution-processed graphene transparent electrodes. ACS Nano **4**(1), 43–48 (2010)
134. Chang, H., et al.: A transparent, flexible, low-temperature, and solution-processible graphene composite electrode. Adv. Func. Mater. **20**(17), 2893–2902 (2010)
135. Wu, J., et al.: light-emitting diodes on solution-processed graphene transparent electrodes. ACS Nano **4**(1), 43–48 (2010)
136. Hwang, J.O., et al.: Workfunction-tunable, N-doped reduced graphene transparent electrodes for high-performance polymer light-emitting diodes. ACS Nano **6**(1), 159–167 (2012)
137. Zhang, Z., et al.: Rosin-enabled ultraclean and damage-free transfer of graphene for large-area flexible organic light-emitting diodes. Nat. Commun. **8**(1), 1–9 (2017)
138. Han, T.-H., et al.: Approaching ultimate flexible organic light-emitting diodes using a graphene anode. NPG Asia Mater. **8**(9), e303–e303 (2016)
139. Lee, J., et al.: Synergetic electrode architecture for efficient graphene-based flexible organic light-emitting diodes. Nat. Commun. **7**(1), 1–9 (2016)
140. Park, I.J., et al.: Flexible and transparent graphene electrode architecture with selective defect decoration for organic light-emitting diodes. Adv. Func. Mater. **28**(10), 1704435 (2018)
141. Kuruvila, A., et al.: Organic light emitting diodes with environmentally and thermally stable doped graphene electrodes. J. Mater. Chem. C **2**(34), 6940–6945 (2014)
142. Kidambi, P.R., et al.: Multifunctional oxides for integrated manufacturing of efficient graphene electrodes for organic electronics. Appl. Phys. Lett. **106**(6), 20_1 (2015)
143. Sanders, S., et al.: Engineering high charge transfer n-doping of graphene electrodes and its application to organic electronics. Nanoscale **7**(30), 13135–13142 (2015)
144. Jia, S., et al.: Graphene oxide/graphene vertical heterostructure electrodes for highly efficient and flexible organic light emitting diodes. Nanoscale **8**(20), 10714–10723 (2016)
145. Yao, L., et al.: Efficient small molecular organic light emitting diode with graphene cathode covered by a Sm layer with nano-hollows and n-doped by Bphen: Cs2CO$_3$ in the hollows. Nanotechnology **28**(10), 105201 (2017)

146. Chen, Y., et al.: Microscale-patterned graphene electrodes for organic light-emitting devices by a simple patterning strategy. Adv. Opt. Mater. **6**(13), 1701348 (2018)
147. Shin, J.-W., et al.: Display process compatible accurate graphene patterning for OLED applications. 2D Mater. **5**(1), 014003 (2017)
148. Tae Lim, J., et al.: Flexion bonding transfer of multilayered graphene as a top electrode in transparent organic light-emitting diodes. Sci. Rep. **5**(1), 1–11 (2015)
149. Priyadarsini, S., et al.: Graphene and graphene oxide as nanomaterials for medicine and biology application. J. Nanostructure Chem. **8**(2), 123–137 (2018)
150. Zhang, J., et al.: Graphene oxide as a matrix for enzyme immobilization. Langmuir **26**(9), 6083–6085 (2010)
151. Shao, Y., et al.: Graphene based electrochemical sensors and biosensors: a review. Electroanal. Int. J. Devoted Fundam. Pract. Asp. Electroanal. **22**(10), 1027–1036 (2010)
152. Srivastava, G., et al.: Functionalized graphene sheets as immobilization matrix for fenugreek β-amylase: enzyme kinetics and stability studies. PLoS One **9**(11), e113408 (2014)
153. Chang, H., et al.: Graphene fluorescence resonance energy transfer aptasensor for the thrombin detection. Anal. Chem. **82**(6), 2341–2346 (2010)
154. Tang, L., et al.: DNA-directed self-assembly of graphene oxide with applications to ultrasensitive oligonucleotide assay. ACS Nano **5**(5), 3817–3822 (2011)
155. Chen, B., et al.: Polyethylenimine-functionalized graphene oxide as an efficient gene delivery vector. J. Mater. Chem. **21**(21), 7736–7741 (2011)
156. Yang, Y., et al.: Construction of a graphene oxide based noncovalent multiple nanosupramolecular assembly as a scaffold for drug delivery. Chem.—A Eur. J. **18**(14), 4208–4215 (2012)
157. Kim, H., et al.: Graphene oxide–polyethylenimine nanoconstruct as a gene delivery vector and bioimaging tool. Bioconjug. Chem. **22**(12), 2558–2567 (2011)
158. Zhou, X., et al.: Ultra-small graphene oxide functionalized with polyethylenimine (PEI) for very efficient gene delivery in cell and zebrafish embryos. Nano Res. **5**(10), 703–709 (2012)
159. Nohwal, B., Pundir, C.: An electrochemical CD59 targeted noninvasive immunosensor based on graphene oxide nanoparticles embodied pencil graphite for detection of lung cancer. Microchem. J. **156**, 104957 (2020)
160. Xu, C., et al.: Encapsulating gold nanoparticles or nanorods in graphene oxide shells as a novel gene vector. ACS Appl. Mater. Interfaces **5**(7), 2715–2724 (2013)
161. Samadian, H., et al.: A de novo theranostic nanomedicine composed of PEGylated graphene oxide and gold nanoparticles for cancer therapy. J. Mater. Res. **35**(4), 430–441 (2020)
162. Dong, X., et al.: Graphene as a novel matrix for the analysis of small molecules by MALDI-TOF MS. Anal. Chem. **82**(14), 6208–6214 (2010)
163. Wang, Y., et al.: Application of graphene-modified electrode for selective detection of dopamine. Electrochem. Commun. **11**(4), 889–892 (2009)
164. Kanakia, S., et al.: Towards an advanced graphene-based magnetic resonance imaging contrast agent: sub-acute toxicity and efficacy studies in small animals. Sci. Rep. **5**(1), 1–12 (2015)
165. Shi, S., et al.: Chelator-free radiolabeling of nanographene: breaking the stereotype of chelation. Angew. Chem. Int. Ed. **56**(11), 2889–2892 (2017)
166. Mathew, T., et al.: Graphene-based functional nanomaterials for biomedical and bioanalysis applications. FlatChem **23**, 100184 (2020)
167. Tadyszak, K., Wychowaniec, J.K., Litowczenko, J.: Biomedical applications of graphene-based structures. Nanomaterials **8**(11), 944 (2018)
168. Fiorillo, M., et al.: Graphene oxide selectively targets cancer stem cells, across multiple tumor types: implications for non-toxic cancer treatment, via "differentiation-based nano-therapy." Oncotarget **6**(6), 3553 (2015)
169. Michaud, M., et al.: Autophagy-dependent anticancer immune responses induced by chemotherapeutic agents in mice. Science **334**(6062), 1573–1577 (2011)
170. de Faria, A.F., et al.: Anti-adhesion and antibacterial activity of silver nanoparticles supported on graphene oxide sheets. Colloids Surf. B **113**, 115–124 (2014)

171. Das, M.R., et al.: Synthesis of silver nanoparticles in an aqueous suspension of graphene oxide sheets and its antimicrobial activity. Colloids Surf. B **83**(1), 16–22 (2011)
172. Bao, Q., Zhang, D., Qi, P.: Synthesis and characterization of silver nanoparticle and graphene oxide nanosheet composites as a bactericidal agent for water disinfection. J. Colloid Interface Sci. **360**(2), 463–470 (2011)
173. GhosalKishor, K., Sarkar, K.: Biomedical applications of graphene nanomaterials and beyond. ACS Biomater. Sci. Eng. **4**, 2653–2703 (2018)
174. Jin, R., et al.: Self-assembled graphene–dextran nanohybrid for killing drug-resistant cancer cells. ACS Appl. Mater. Interfaces **5**(15), 7181–7189 (2013)
175. Zhou, K., et al.: One-pot preparation of graphene/Fe_3O_4 composites by a solvothermal reaction. New J. Chem. **34**(12), 2950–2955 (2010)
176. Dong, J., et al.: Application of graphene quantum dots for simultaneous fluorescence imaging and tumor-targeted drug delivery. Sens. Actuators B Chem. **256**, 616–623 (2018)
177. Liu, Z., et al.: PEGylated nanographene oxide for delivery of water-insoluble cancer drugs. J. Am. Chem. Soc. **130**(33), 10876–10877 (2008)
178. Wang, C., et al.: Gold nanoclusters and graphene nanocomposites for drug delivery and imaging of cancer cells. Angew. Chem. **123**(49), 11848–11852 (2011)
179. Zhang, L., et al.: Functional graphene oxide as a nanocarrier for controlled loading and targeted delivery of mixed anticancer drugs. Small **6**(4), 537–544 (2010)
180. Zhang, Y., et al.: A graphene modified anode to improve the performance of microbial fuel cells. J. Power Sources **196**(13), 5402–5407 (2011)
181. Hu, H., et al.: Engineering of a novel pluronic F127/graphene nanohybrid for pH responsive drug delivery. J. Biomed. Mater. Res. Part A **100**(1), 141–148 (2012)
182. Wen, H., et al.: Engineered redox-responsive PEG detachment mechanism in PEGylated nano-graphene oxide for intracellular drug delivery. Small **8**(5), 760–769 (2012)
183. Pan, Y., et al.: Water-soluble poly (N-isopropylacrylamide)–graphene sheets synthesized via click chemistry for drug delivery. Adv. Func. Mater. **21**(14), 2754–2763 (2011)
184. Zhang, L., et al.: High strength graphene oxide/polyvinyl alcohol composite hydrogels. J. Mater. Chem. **21**(28), 10399–10406 (2011)
185. Chen, G.-Y., et al.: A graphene-based platform for induced pluripotent stem cells culture and differentiation. Biomaterials **33**(2), 418–427 (2012)
186. Verre, A.F., et al.: Improving the glial differentiation of human Schwann-like adipose-derived stem cells with graphene oxide substrates. Interface Focus **8**(3), 20180002 (2018)
187. Kim, J., et al.: Bioactive effects of graphene oxide cell culture substratum on structure and function of human adipose-derived stem cells. J. Biomed. Mater. Res. Part A Off. J. Soc. Biomater., Jpn. Soc. Biomater. Aust. Soc. Biomater. Korean Soc. Biomater. **101**(12), 3520–3530 (2013)
188. Li, N., et al.: Three-dimensional graphene foam as a biocompatible and conductive scaffold for neural stem cells. Sci. Rep. **3**(1), 1–6 (2013)
189. Shin, S.R., et al.: Cell-laden microengineered and mechanically tunable hybrid hydrogels of gelatin and graphene oxide. Adv. Mater. **25**(44), 6385–6391 (2013)
190. Roy, R.K., Lee, K.R.: Biomedical applications of diamond-like carbon coatings: a review. J. Biomed. Mater. Res. Part B: Appl. Biomater. Off. J. Soc. Biomater. Jpn. Soc. Biomater. Aust. Soc. Biomater. Korean Soc. Biomater. **83**(1), 72–84 (2007)
191. Mohandes, F., Salavati-Niasari, M.: In vitro comparative study of pure hydroxyapatite nanorods and novel polyethylene glycol/graphene oxide/hydroxyapatite nanocomposite. J. Nanopart. Res. **16**(9), 1–12 (2014)
192. Mohandes, F., Salavati-Niasari, M.: Freeze-drying synthesis, characterization and in vitro bioactivity of chitosan/graphene oxide/hydroxyapatite nanocomposite. RSC Adv. **4**(49), 25993–26001 (2014)
193. Lee, J.H., et al.: Reduced graphene oxide-coated hydroxyapatite composites stimulate spontaneous osteogenic differentiation of human mesenchymal stem cells. Nanoscale **7**(27), 11642–11651 (2015)

194. Lingamdinne, L.P., Koduru, J.R., Karri, R.R.: A comprehensive review of applications of magnetic graphene oxide based nanocomposites for sustainable water purification. J. Environ. Manage. **231**, 622–634 (2019)
195. Li, B., et al.: Superparamagnetic Fe_3O_4 nanocrystals@ graphene composites for energy storage devices. J. Mater. Chem. **21**(13), 5069–5075 (2011)
196. Chandra, V., et al.: Water-dispersible magnetite-reduced graphene oxide composites for arsenic removal. ACS Nano **4**(7), 3979–3986 (2010)
197. Luo, X., et al.: Adsorption of As (III) and As (V) from water using magnetite Fe_3O_4-reduced graphite oxide–MnO_2 nanocomposites. Chem. Eng. J. **187**, 45–52 (2012)
198. Lingamdinne, L.P., et al.: Porous graphene oxide based inverse spinel nickel ferrite nanocomposites for the enhanced adsorption removal of arsenic. RSC Adv. **6**(77), 73776–73789 (2016)
199. Bhunia, P., et al.: A strategically designed porous iron–iron oxide matrix on graphene for heavy metal adsorption. Chem. Commun. **48**(79), 9888–9890 (2012)
200. Wang, H., et al.: Facile synthesis of polypyrrole decorated reduced graphene oxide–Fe_3O_4 magnetic composites and its application for the Cr (VI) removal. Chem. Eng. J. **262**, 597–606 (2015)
201. Sun, Y., et al.: Interaction between Eu (III) and graphene oxide nanosheets investigated by batch and extended X-ray absorption fine structure spectroscopy and by modeling techniques. Environ. Sci. Technol. **46**(11), 6020–6027 (2012)
202. Zong, P., et al.: Synthesis and application of magnetic graphene/iron oxides composite for the removal of U (VI) from aqueous solutions. Chem. Eng. J. **220**, 45–52 (2013)
203. Zhao, Y., et al.: Efficient enrichment of uranium (VI) on amidoximated magnetite/graphene oxide composites. RSC Adv. **3**(41), 18952–18959 (2013)
204. Tan, L., et al.: The synthesis of a manganese dioxide–iron oxide–graphene magnetic nanocomposite for enhanced uranium (VI) removal. New J. Chem. **39**(2), 868–876 (2015)
205. Fu, J., et al.: Adsorption of methylene blue by a high-efficiency adsorbent (polydopamine microspheres): kinetics, isotherm, thermodynamics and mechanism analysis. Chem. Eng. J. **259**, 53–61 (2015)
206. Singh, K.P., et al.: Optimizing adsorption of crystal violet dye from water by magnetic nanocomposite using response surface modeling approach. J. Hazard. Mater. **186**(2–3), 1462–1473 (2011)
207. Yang, Z., et al.: Magnetic nanomaterial derived from graphene oxide/layered double hydroxide hybrid for efficient removal of methyl orange from aqueous solution. J. Colloid Interface Sci. **408**, 25–32 (2013)
208. Abdi, G., et al.: Removal of dye and heavy metal ion using a novel synthetic polyethersulfone nanofiltration membrane modified by magnetic graphene oxide/metformin hybrid. J. Membr. Sci. **552**, 326–335 (2018)
209. Khurana, I., et al.: Batch and dynamic adsorption of Eriochrome Black T from water on magnetic graphene oxide: experimental and theoretical studies. J. Environ. Chem. Eng. **6**(1), 468–477 (2018)
210. Liu, G., et al.: Preparation of a magnetic molecularly imprinted polymer using g-C_3N_4–Fe_3O_4 for atrazine adsorption. Mater. Lett. **160**, 472–475 (2015)
211. Boruah, P.K., et al.: Magnetically recoverable Fe_3O_4/graphene nanocomposite towards efficient removal of triazine pesticides from aqueous solution: investigation of the adsorption phenomenon and specific ion effect. Chemosphere **168**, 1058–1067 (2017)

Graphene Nanotechnology for Renewable Energy Systems

M. Krishna Kumar and M. Muthuvinayagam

Abstract The high degree of mechanical, electrical, and thermal conductivity of graphene enables its application in the renewable energy sector. Graphene plays a vital role in diodes, photovoltaic cells, supercapacitors, batteries, and full cells applications and it enhances the existing efficiency in a tremendous way. The addition of graphene can do tremendous effects on all kinds of inorganic and organic materials for solar cells. In the supercapacitor domain, graphene plays a vital role to provide a large surface which can be useful for energy storage applications. The hybrid supercapacitor and battery trends in research and technology have been discussed. The electrical vehicles and green applications need and accomplishments of graphene and its derivatives have also been described.

Keywords Solar energy · Fuel cells · Graphene · Graphene oxide (GO) · Metal oxide nanocomposites

1 Introduction

Renewable energy systems implementations are essential due to the shortfall in fossil fuels and resources. The available methods of renewable energy systems fall under solar, wind, tidal, hydro, biomass and etc. A major resource on that is solar and wind energy. The effective utilization of the same is to be ensured for various applications. Graphene and its derivative compounds are more useful for energy storage applications. A single layer of carbon atoms arranged in such a honeycomb structure forms a single graphene sheet. Several sheets stacked one on top of the other are regarded as multilayer graphene, up to the point where the material becomes graphite (Fig. 1).

M. K. Kumar (✉)
Department of Physics and Electronics, CHRIST (Deemed to be University), Bengaluru 560 029, Karnataka, India
e-mail: krishnakumar.m@christuniversity.in

M. Muthuvinayagam
Department of Physical Sciences, Saveetha School of Engineering, Saveetha University (SIMATS), Chennai 600 077, Tamil Nadu, India

© The Author(s), under exclusive license to Springer Nature Singapore Pte Ltd. 2023
R. T. Subramaniam et al. (eds.), *Graphene*, Engineering Materials,
https://doi.org/10.1007/978-981-99-1206-3_8

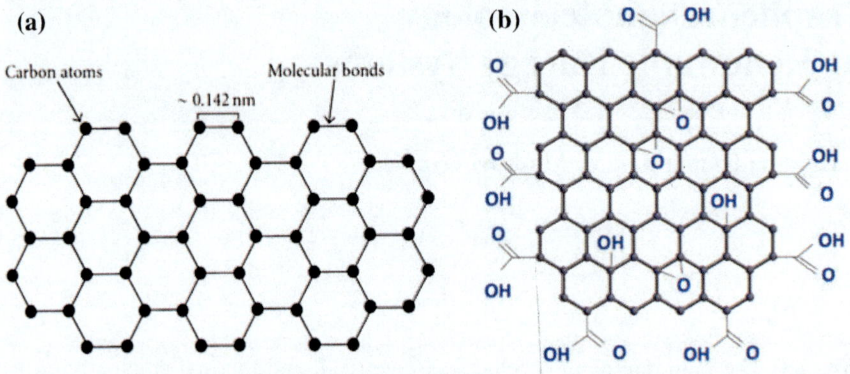

Fig. 1 **a** Structure of pristine graphene in a honeycomb structure, **b** graphene oxide. Reproduced with permission from Tiwari et al. [1] under CC BY 4.0

This book chapter is going to discuss the applications of graphene for various renewable energy storage systems applications. The carbon atoms are arranged in a single layer, thus, resulting in the formation of graphene which has been the keen interest due to its two-dimensional arrangement. Graphene system consists of a single-atom thick hexagonal honeycomb lattice and a unique class of material with a wide range of desirable properties [1]. It is an allotrope of carbon in the form of a plane of sp^2 with a molecular bond with length of 0.142 nm. The layers of Graphene when stacked on top of each other, it forms graphite, with an "inter-planar spacing of 0.335 nm" [1]. The separate layers of graphene in graphite are held together by van der Waals forces, which can be overcome during the exfoliation of graphene from graphite. Graphene has high degree of charge carrier mobility, which means that the electrons move quickly through the material. It is also considered to be one of the most important properties to consider it as a very good material. Charge carriers can have high mobility of 15,000 cm^2 V^{-1} s^{-1} at ambient conditions and can be continually switched between electrons and holes in concentrations of 1013 cm^{-2} [2]. The bandgap of the graphene is zero and graphene oxide is 0.9 eV. Hence pristine graphene is not a semiconductor and it has to be tuned with bandgap for the semiconducting applications as shown in Fig. 1 [3, 4]. The mobility of graphene has a consistently high value despite the high value in electrically and chemically doped devices. In addition, the electron scatter is the least in graphene as compared to other materials [5]. Electron–electron scattering in monolayer graphene at finite "n" carrier concentration can strongly be suppressed if a metallic gate bulk conductor is placed at "d" distance from a graphene layer of about 1 nm. This "close-gate" regime has become accessible due to the use of van der Waals assembly that allows atomically sharp interfaces and ultra-thin electrostatically. Also, it has a high mechanical strength of about 100–300 times stronger than steel and has a tensile strength of 130 GPA and Young's Modulus of 1 TPA [6]. Hence it can be used without any hesitation in many fields for long life of the technology. This versatile substance has a capability to revolutionize a number of technologically important areas including the energy sector [2, 6, 7]. In short, graphene-based solar cells, graphene photovoltaic cells, green energy harvesting diodes, perovskites solar cells, graphene supercapacitors, graphene batteries, and graphene fuel cell technologies could really benefit the renewable energy sector in the near future.

2 Solar Technology

Solar energy is the major alternate energy resource for fossil fuels, since all other renewable sources are indirect forms of solar energy. Solar panels can work best during periods of strong sunlight, but start to wane when it gets cloudy or rainy. Our dependence upon renewable energy becomes more apparent, so the need for efficient solar cells becomes more crucial, especially when they are one of the easiest and cheapest ways to generate clean energy. For instance, graphene-based various types of solar cells have been researched [8]. A breakthrough in graphene-based solar panels could change all that, by allowing solar panels to generate electricity during inclement weather [9]. Graphene-based solar cells would be able to derive energy from raindrops that happen to fall on the panel, by taking advantage of the various salts present within the liquid. The graphene sheets that make up the solar cells would be able to separate the positively charged ions in rainwater [10]. These positive ions then bind to the ultra-thin layer of the graphene to form a double layer with the presence of electrons. Graphene as a material is both strong and light, and can hold energy better than graphite. Graphene could have an important role to play in anti-reflection coatings for solar cells. It is being developed into anti-reflection coatings, so the integration of graphene into solar applications [11]. It has lower reflectance near the ultraviolet part of the spectrum from 35 to 15%. However, the results can accelerate the design of future solar cells, and open up thinking about alternative electricity generating capabilities for solar cells. This is of course in addition to current trends toward the wider use of solar panels, such as integrating them into roofs, walls and windows of buildings, as well as the movement to provide decentralized storage of power generated by solar panels via battery systems. The improvement of perovskites solar cells (PSCs) has been a highly promising next-generation solar power source with very high efficiency. Arora et al. [12] added a reduced graphene oxide (GO) spacer layer to a PSC resulting in low-cost production of PSCs with 20% efficiency, retained up to 95% after 1000 h of operation. Hence, the presence of graphene will have physical effects on solar cells by its different kinds of integration in various kinds of solar cells, such as the effects of doped graphene in Solar Cells [13], graphene-Silicon Solar Cells [14], graphene-Polymer Solar Cells [15], Dye-Sensitized Solar Cells (DSSC's) [16], graphene-Quantum Dot (QD) Solar Cells [17], graphene-Tandem Solar Cells [18], graphene-Perovskites Solar Cells (PSCs) [19], graphene-Organic Solar Cells [20], graphene Bulk-Heterojunction Solar Cells [21], graphene Transparent Electrodes [22], graphene Photoactive Layers [23], graphene Schottky Junction GaAS Solar Cells [24], respectively. A brief information about the various kinds of renewable energy technologies using graphene has been discussed

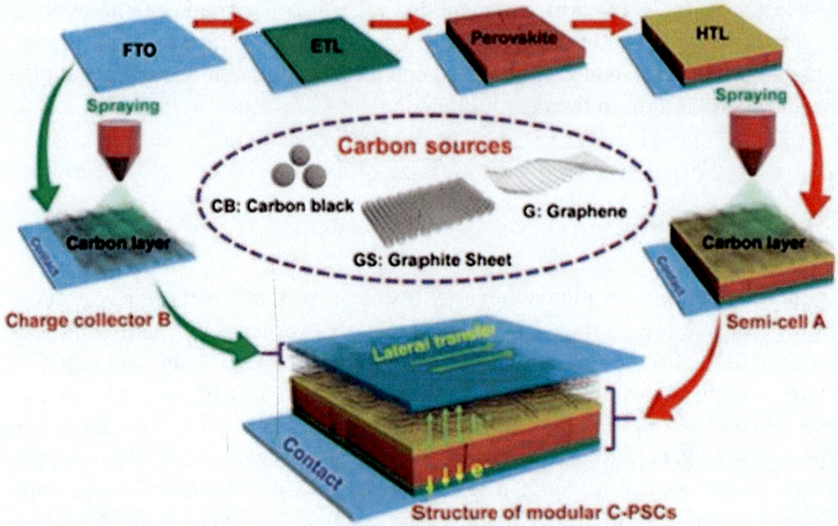

Fig. 2 Fabrication process for the modular C-PSCs. Reproduced with permission from Zhang et al. [25]

in the following subsections. To incorporate graphene and its derivatives into photoelectrical devices, the transfer and assembly methods are necessary. Recently, Shi et al. proposed an innovative modular carbon-based perovskites solar cell (C-PSC) [25]. It is shown in Fig. 2.

The ideal transparent electrode for solar cell, the requirements of high transparency, low sheet resistance, robust chemical stability, and low cost should be simultaneously fulfilled. In 2015, You et al. [26] fabricated a semi-transparent perovskites solar cell by laminating multilayer graphene as top transparent electrodes. The conductivity was further improved with the help of poly-(3,4-ethylenedioxythiophene):poly(styrenesulfonate) (PEDOT:PSS) thin film thickness of about $\approx$20 nm). Hadadian et al. [27] incorporated nitrogen-doped RGO (N-RGO) into mixed organic–inorganic halide perovskite. The resulting device exhibits an almost hysteresis-free PCE of 18.7%.

2.1 Graphene Solar Cell—Principle

The principle of a graphene-based solar cell is similar to present inorganic/silicon solar cells, the exception is materials are replaced with graphene derivatives. It has to add in any device to improve the parameters that can increase the operational efficiency. In the graphene-based solar cells, the two essential parameters potentially changes the nature of the device, they are "number of graphene layers" in the device and the "effects of doping a graphene-based" material [28].

2.2 Effects of Graphene Layers in Solar Cells

The solar photovoltaic system shall have the optical transparency on the glass sheet which allows electromagnetic sunlight. The number of graphene layers can be characterized by a relationship between optical transparency and sheet resistance by a proportional decrease. The optimization of directly grown, continuous layers of graphene can be done to achieve superior performance. Graphene has a zero band gap behaving as a metal and makes a Schottky junction in combination with semiconductors [29]. Thus, with different thicknesses of graphene on bare silicon by controlling the graphene thickness, the work function can be significantly improved, the open circuit voltage is increased, and the energy conversion efficiency also enhances [29]. Single layer graphene shows an optical transparency of about 97.7%. The three-layer graphene exhibits around 90.8% optical transparency, and the addition of each layer corresponds to a 2.3% decrease in optical transparency. The single sheet of graphene produces a sheet resistance of about 2.1 $k\Omega sq^{-1}$ and 350 Ωsq^{-1}, while retaining 90% optical transparency [30]. The effect of quenching multiple graphene layers can be up to 11% higher efficiency than monolayer graphene addition, due to a higher hole accepting density of state [31, 32].

2.3 Effects of Doped Graphene in Solar Cells

The doping of heteroatom onto a sheet of graphene significantly alters the chemical, electronic, physical, and photonic properties as shown in Fig. 3. There are two main types of doping- p-type and n-type. P-type doping utilizes trivalent atoms, such as boron, which extracts an electron off the graphene sheet and creates a hole, a process known as hole doping, where the hole is created in the valence band of the graphene sheet. On the other hand, n-type doping involves pentavalent atoms, such as phosphorous, and is an electron donating doping approach that facilitates a free electron from the pentavalent atom onto the graphene sheet. The free electron in this instance is facilitated in the conductance band of the graphene sheet. This doping in to the graphene sheet can be done through various methods, such as chemical doping, ball milling, thermal annealing, chemical vapor deposition (CVD) plasma treatment, etc., [33]. Both doping process improves efficiency of the solar cell. Seong et al. [13] investigated the Graphene (GR), GR/N-doped TiO_2 (GNT) composite was prepared to both expand absorption into the visible light region and to enhance the electron-transfer rate; the material was used as a photoelectrode in our DSSC system, which had a higher maximum power conversion efficiency than that of plain TiO_2 photoelectrode. Sangwoo et al. [34] studied the p-type cationic nitrogen doped graphene for altering the properties. Priyadarsini et al. [35] researched the effects of doping of Nitrogen, Boron, Phosphorous, and Sulfur atoms on graphene and found the effects on it.

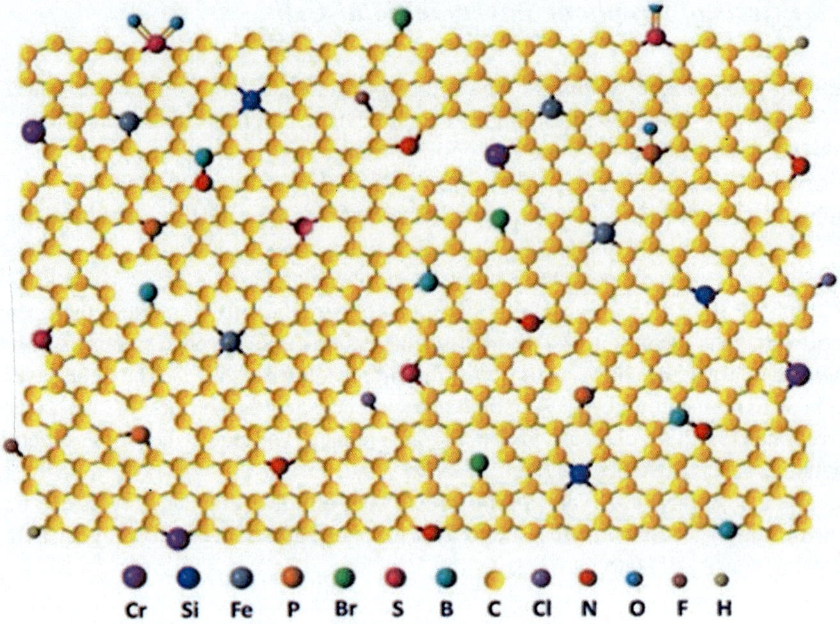

Fig. 3 Schematic diagram of possible hetero atom doped graphene

2.4 *Graphene-Silicon Solar Cells*

Graphene-based films for solar cells can be produced with a pre-determined thickness and complete coverage. Pure silicon cells' performance is superior already; hence Graphene-silicon solar cells are to be researched to enhance its efficiency by dopants amount used. Schottky junction solar cells can be fabricated by directly depositing a thin layer of metal, transparency electrode on a moderately doped semiconductor wafer [36]. Graphene can be added by various junctions in Graphene-silicon solar cells, such as p-type heterojunctions, n-type heterojunctions and Schottky junctions. The tunability of Graphene is a promising approach for hybrid solar cells and it is just a matter of time until their efficiency is larger than pure silicon cells. N-type heterojunctions can generate a 0.55–0.57 V to facilitate electron-hole separation and Schottky junctions have a power conversion efficiency (PCE) of 5.67% [36, 37]. Due to the excellent optical and electrical properties of graphene, there is a great interest in developing Graphene/Si Schottky junction solar cells [38].

2.5 Graphene-Based Polymer Solar Cells

Polymeric materials offer great advantages than inorganic counterparts due to its tunability, low cost and simple fabrication processes. Graphene has a great potential in transparent electrodes as a replacement for indium tin oxide (ITO) in polymer-based solar cells [39]. Replacing ITO with monolayer graphene in organic solar cells yields comparable performance due to the increased optical absorption. Graphene can be mixed with polymeric material to produce a material with a band gap of up to 3.6 eV which prohibits an electron migration from the cathode to the anode which stops current leakage and charges recombination. The organic solar cells with four-layer graphene can attain at least 92% of the same organic photovoltaic device with an optimized ITO electrode. The hybrid materials possess a better energetic relationship, as the Fermi-level of the graphene and the semiconducting layer are closer together for an efficient charge injection [40, 41]. Graphene-polymer transparent electrodes possess a high work function and conductivity, but it does have a limit of 65% light transmittance [42]. In addition to reducing the graphene into hybrids, CVD-produced graphene can also be used as transparent electrodes. The oxygen functional groups usually improve the open circuit voltage, but conductivity will be reduced due to the sp^2 hybridized covalent network being disrupted by sp^3 bonds around the functionalized carbons. Non-covalent functionalized CVD-grown graphene have a good conductivity and can have up to 0.55 V open circuit voltage, a fill factor of 55% and a PCE of 1.71%. Flexibility of graphene allows the solar cell to bend up to 78° more than pure ITO electrodes [43]. The large surface area of graphene allows for a continuous pathway and multiple donor/acceptor sites for efficient electron transfer.

2.6 Dye-Sensitized Solar Cells (DSSCs)

The dye-sensitized solar cell is a renewable energy technology that gained the attention of the photovoltaic community and DSSCs are different when compared to other types of solar cells [44]. Muchuweni et al. [16] studied the ground and excited state of the graphene-based active material DSSC will do the electronic conduction as shown in Fig. 4. They contain a semiconducting material with a photo-sensitive dye as the anode coupled with a pure metal cathode and an electrolyte solution. The sensitizers are usually designed to have functional groups such as –COOH, $-PO_3H_2$ and $-B(OH)_2$ for stable adsorption onto the semiconductor substrate. A thin layer of oxide semiconducting materials such as TiO_2, Nb_2O_5, ZnO, SnO_2 (n-type), and NiO (p-type) on a transparent conducting glass plate made up of fluorine-doped tin oxide (FTO) or Indium tin oxide (ITO). These oxides have a wide energy band gap of 3–3.2 eV [45]. The graphene addition can increase the loading efficiency of the dye molecules, increase the interfacial area and improve the conductivity of the electrons to compete against the effects of charge recombination. Compared with traditional

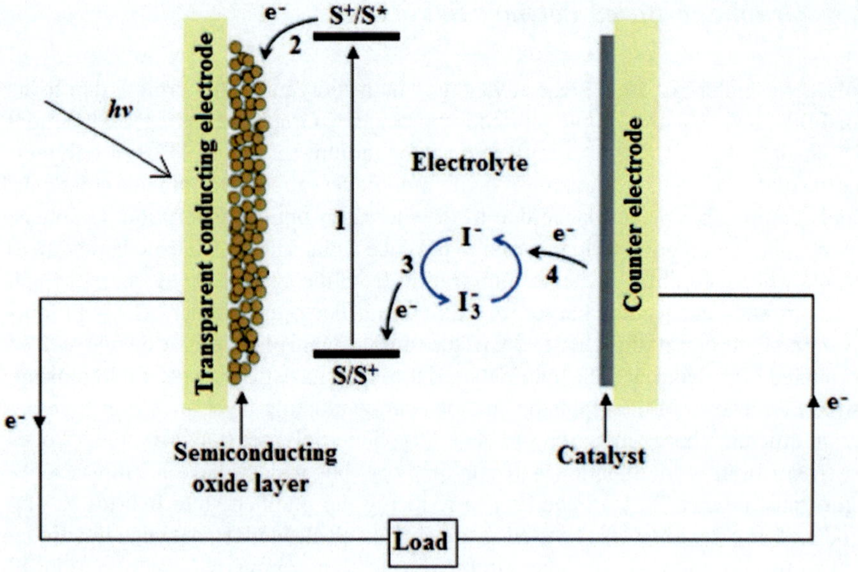

Fig. 4 Schematic diagram of the basic operation principle of a typical DSSC. Reproduced with permission from Muchuweni et al. [16] under CC BY-NC 3.0. (S/S*—Ground state/Excited state; S^+ is the oxidized dye molecule)

DSSCs, cells with graphene photoanodes have indeed demonstrated improved performance. The incorporation of graphene into DSSCs improves the light scattering at the photoanode, efficiently disperses the dye molecules and provides greater efficiency than pure electrodes [46]. Dhonde et al. [47] demonstrated that an optimal loading of Cu and graphene can boost the power conversion efficiency (PCE) of the DSSC (9.81%) by 47% higher than undoped DSSC (6.66%). Kuan et al. [48] researched that, B–N co-doped graphene (BNG) exhibits greater advantages in both conductivity and stability, which is an important aspect of the counter electrode (CE) in DSSCs.

2.7 *Graphene Quantum Dot (GQD) Solar Cells*

GQDs have been constructed by one or only a few graphene sheets which are less than 10 nm in size, thus making this material to be a promising candidate with the excellent physical and electrical properties of graphene [49, 50]. The properties are to be adjusted by incorporating heteroatom doped GQDs such as B, N, F and S [51]. Both graphene and Carbon Nanotubes (CNTs) have been hybridized with quantum dots to make functioning GQDs solar cells. The Graphene/CdS bilayer GQD configuration have an efficiency of up to 16%, which outperforms Carbon Nanotubes/CdS by 7, and 11% for other carbon allotropes. This attributes to graphene producing a better scaffold to incorporate the quantum dots; the layered structure provides a fast electron transfer from the QD to the graphene while suppressing the recombination of charges [52].

2.8 Graphene-Based Perovskites Solar Cells (PSCs)

Perovskites solar cells (PSCs) have made great strides over the last few years due to their interesting bandgap and absorption properties that produce high PCEs. Perovskites solar cells have a standard structure, including the type of materials that are used, so the substitution of one material for another is a relatively simple process that leads to highly tuneable solar cell devices [12]. The best results in these solar cells can be achieved when the nanocomposites are utilized as an n-type electron collection layer. The graphene is present as a monolayer and is only present as 0.6 wt.% of the whole cell. Any amount above this drops the efficiency and is dependent upon a thin collecting layer. These cells also possess short circuit and open circuit values of 12–21.9 $mAcm^{-2}$ and 1.05 V. When paired with an efficient light absorber, GO can be used as a hole conductor for inverted solar cells. PSCs have the prominent advantages of flexibility, low-temperature solution processing, low cost, and large-area production. It is worth noting that the PSC has achieved a PCE of ~16.0% for the modulus [19]. The fabrication of this class of PSC is more complex but provides a PCE greater than similar solar cells without the GO layer. Thinner layers of GO (2 nm) can produce higher efficiencies. The average short and open circuit values in these solar cells are around 15.58 $mAcm^{-2}$ and 0.99 V, respectively. The higher stability is attributed to the increased resistance that reduced graphene oxides possess against oxygen and moisture compared to other graphene derivatives. Redondo et al. [53] added very low concentrations of few-layer graphene platelets (up to 24×10^{-3}wt %) with $MAPbI_3$ films and fabricated solar cells. The lowest graphene content delays the degradation of films with time and light irradiation and leads to enhanced photovoltaic performance and stability of the solar cells, with relative improvement over devices without graphene of 15% in the power conversion efficiency and PCE. Paolo et al. [54] have done work on Carbon perovskite solar cells (C-PSCs), using carbon-based counter electrodes (C-CEs) and found the solution to instability issues and C-PSCs achieved a maximum PCE of 15.81%.

An example of perovskites solar cell is shown in Fig. 5 studied by O'Keeffe et al. [55]. The hole transport layer (HTL) is the combination of spiro-OMeTAD and the electron transport layer (ETL) made by PSC-G (m-$TiO2^+$ graphene + perovskites). Electron Transport Layer (ETL) is PSC-NoG: m-$TiO2^+$ perovskite. C-TiO2 is a compact TiO_2 hole-blocking layer. FTO: fluorine-doped tin oxide. The capping layer is composed of large-sized perovskite crystals with dimensions of about 500 nm. The ETL is composed of small-sized perovskite crystals with dimensions of 20–40 nm, spatially restrained within the m-TiO_2 layer. Indeed PSC-NoG presents a greater reduction of power conversion efficiency with respect to PSC-G, retaining

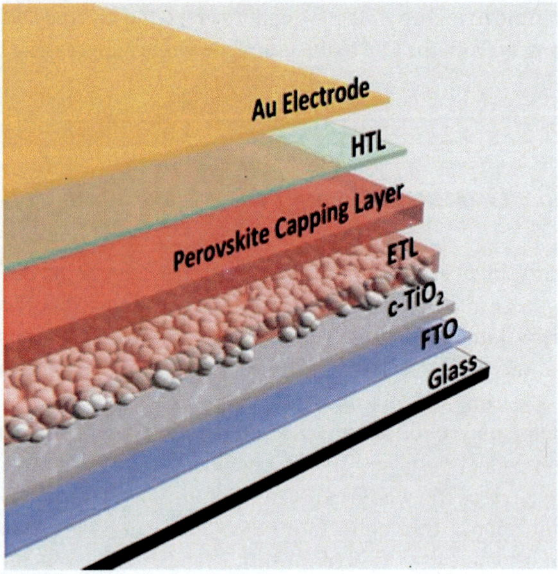

Fig. 5 Schematic representation of the layer structure of the example PSCs. Reproduced with permission from O'keeffe et al. [55]

93% (reduction of 7%) of the initial PCE after 1 week of aging while PSC-NoG only 86% (reduction of 14%).

2.9 *Graphene- Multi-junction Solar Cells*

Multi-junction solar cells otherwise known as Tandem solar cells composed of two or more sub-cells that are stacked in either a series or parallel configuration. A single solar cell can produce up to 40% solar energy conversion efficiency, but tandem solar cells have the potential to reach up to 86% efficiency. The power conversion efficiency (PCE) of many solar cells has been enhanced to date by employing tandem arrangements [56]. Strategy to raise the efficiency of solar cells is stacking solar cell materials with different band gaps to absorb different colrs of the solar spectrum. There have been some promising developments using graphene oxide based tandem solar cells but as a relatively new area, they show great potential; especially as non-graphene tandem solar cells show relatively high PCEs. One such development is that of graphene oxide and polymer tandem solar cells that consist of 2 sub-cells. The cells consist of a bilayer of Cs-neutralized graphene oxide and pure graphene oxide connected by a charge recombinant layer of molybdenum trioxide (MoO_3) and Anrango-Camacho et al. [57]. The world-record efficiency of a 3 Junction cell comprises of lattice-mismatched GaInP, 1.42 eV-InGaAs and 1.0 eV-InGaAs, which can reach 44.4% efficiency under 302 suns, whereas a four-junction (4 J) GaInP/GaAs; GaInAsP/ GaInAs could reach 46.0% at 508 suns [38]. The open circuit voltage of the tandem cell can vary between 1.23 and 1.69 V, but is dependent on

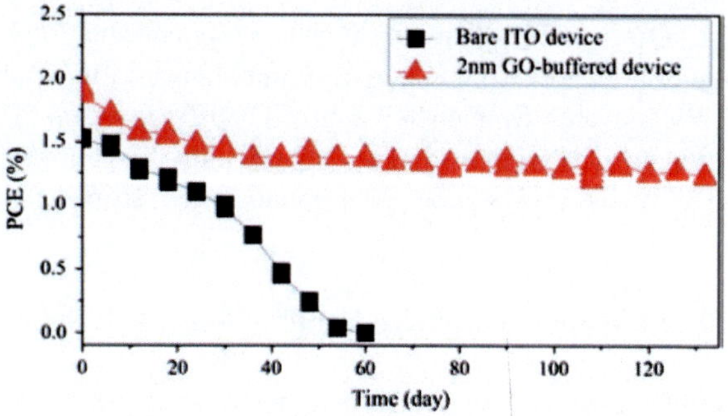

Fig. 6 Variation of PCE responses of an encapsulated ITO-based device and GO-buffered device. Reproduced with permission from Singh and Nalwa [21] under CC BY 3.0

the resistance of the interconnecting layer between the graphene oxide sheets, which is a function of the thickness of the layer. Combinations of these materials have also been used as the hole transport and interconnecting layers for ITO-based sub-cells. The resulting solar cells possess a higher PCE than solar cells that contain the same sub-cells but lack the graphene connecting layers. Even though graphene is not directly involved in the sub-cells, the presence of graphene in the device increases its efficiency. Power conversion efficiency in the range of 10–15% for graphene and inorganic semiconductor-based hybrid heterojunction solar cells, and 15.6% for graphene-containing perovskite solar cells has been observed [21]. The efficiency variations on ITO and GO added ITO is shown in Fig. 6 for understanding.

2.9.1 Graphene-Organic Solar Cells

The light-harvesting organic solar cells (OSCs) are involving different inorganic components; the organic components of the solar cell also play a major role. Organic and inorganic components in a solar cell have advantages and disadvantages, but the optimization of the organic components can produce a more efficient solar cell. Components that might traditionally be inorganic in nature are now being replaced with inorganic-organic hybrid materials that offer greater physical properties, solution processability, cost-effective production, a large surface area, and are much lighter in nature. One concern with many solar cells is the environmental stability, but organic molecules can provide stability against temperature, moisture and chemical degradation in solar cells, even when present as a hybrid material. Traditional organic solar cells contain an active PEDOT:PSS layer and a donor-acceptor blend layer-commonly composed of P3HT or fullerenes. In recent years, the active PEDOT:PSS layer has been replaced by graphene derivatives and are generally used as hole transport layers in organic solar cells [57]. The overall degradation of organic solar cells is a rather complicated process, but to include a few important factors, it involves photo-oxidation of conjugated polymeric materials and fullerene derivatives, eroding of the ITO electrodes, loss of electrical and mechanical contacts, and morphological inhomogeneities. To overcome these drawbacks of PEDOT:PSS-based solar cells, novel inorganic and organic materials have been used in HTLs in BHJ solar cells

[21]. Qing et al. [58] investigated the graphene and graphene oxide (GO) applied flexible organic electronic devices with enhanced efficiency of polymeric photovoltaic (OPV) devices. Spin coated 2 nm GO buffer layer on ITO have the power conversion efficiency (PCE) of a standard copper phthalocyanine (CuPc)/fullerene (C60) based OPV device shows about 30% enhancement from 1.5 to 1.9%.

2.9.2 Graphene Bulk-Heterojunction (BHJ) Solar Cells

Graphene-based solar cells have demonstrated a short span of time, more stability under ambient atmospheric conditions. High electronic conductivity, transparency and flexibility make them useful in heterojunctions in solar cells, where they can be employed in many different ways of electrodes, acceptor layers, donor layers, buffer layers and active layers. BHJ solar cells based on organic small molecules and polymers are the focus of increasing attention [21]. In organic photovoltaic devices, a conjugated polymer layer is used as the donor, while a fullerene-based derivative is used as the acceptor [59]. Poly(3,4- ethylenedioxythiophene):poly(styrene sulfonate) (PEDOT:PSS) is one of the most common interfacial materials used for organic BHJ solar cells [60–62]. However, the multi-junction within the solar cell relies heavily on graphene's specific tuneable parameters, including the thickness, thermal annealing temperature, the concentration of doping on the sheet and its photovoltaic performance. There are many variations of heterojunction solar cells and how graphene derivatives can be incorporated into them, including as transparent electrode, photoactive layers and Gallium Arsenide (GaAs) solar cells [21].

2.9.3 Transparent Graphene Electrodes

Graphene can easily be incorporated into certain layers. Coupled with its excellent electrical, optical, mechanical, and thermal properties, this has allowed graphene to be studied as transparent electrodes in solar cells. Prior to graphene being employed as a transparent electrode, ITO was the most used material because of its high optical transparency. However, ITO is not cost-effective, is brittle, and lacks mechanical flexibility. Graphene exhibits a high optical transparency of 90–100% and a low sheet resistance, even in multiple layered graphene stacks both of which are great properties for transparent electrode applications. There have been many cases of graphene derivatives being employed as both the anode and the cathode in heterojunction solar cells [21]. Some of the common molecules used in these graphene derivatives including polyethylene naphthalate (PEN), PEDOT: PSS, MoO_3 and ZnO. PEDOT:PSS layers are the most common in graphene transparent electrodes, with other materials being incorporated to improve or tune the properties. Electrochemically produced graphene electrodes do have viable figures of merit, but they are still at least an order of magnitude away from being a viable ITO replacement. Graphene electrodes showed average PCEs up to 12.02% and 11.65% for the devices illuminated from FTO and graphene electrodes [19]. In comparison with flexible ITO/PEN electrodes, MoO_3-decorated graphene (Gr-Mo)/PEN electrodes enabled with PSC devices exhibit better mechanical stability, and only 12% degradation in the PCE occurred in graphene-based flexible PSCs after 5000 bending cycles at bending radius of 4 mm. These solar cells have also been found to exhibit a short-circuit photocurrent density up to 14.8 $mAcm^{-2}$, an open circuit voltage up to 0.71 V, with the potential to obtain a form factor of up to 57.6% after 100 tensile flexing cycles [63].

Solar cells from these composite dual graphene electrodes have been found to show no loss of activity under mechanical tension binding tests, a high efficiency and excellent mechanical strength. The doping of the graphene electrode changes the wetting in the PEDOT:PSS layer on the surface of the graphene sheet. The change in properties leads to an enhanced PCE performance across the whole cell. Solar cells of this variety have been produced using 1–3 layers of graphene sheets. Depending on the thickness of the hole transport layer, the PCE of these solar cells can range from 0.71–0.31%. Other area of graphene transparent electrodes research explores Zinc-Graphene anodes. The hybrid photoanode, based around a P3HT, ZnO and ZnS core-shell nanorod array, suspended on a reduced graphene oxide film has been modified with ITO. Each component has a specific various multilayer electrodes based around graphene, gold, P3HT, PCBM, PEDOT:PSS, copper, and PMMA in various compositions have been employed with significantly different results. The transparency of these electrodes varies from 82.3 to 90% with sheet resistances varying massively between 92 and 374 Ωm^{-2}. The PCE of such electrodes can vary between 1.17 and 13.3% [64].

2.9.4 Graphene Photoactive Layers

Graphene and GO can also be employed in heterojunction solar cells as photoactive layers in the form of an active interfacial layer, electron-hole separation layer, and holetransport layer or as an electron-transport layer. Graphene photoactive layers can exhibit a PCE from anywhere between 0.4 and 10.3% depending on the graphene derivative and the type of photoactive layer being produced. There are currently hundreds of graphene photoactive layers being employed as heterojunction solar cells [21, 29, 42]. Photoactive layers are composed of a few layers of pure graphene films and they can be employed as n-type heterojunctions by utilizing n-type silicon alongside the graphene layers. Doping with nitric acid increases the PCE of these pure graphene heterojunctions up to 4.35%, where up to 4.18% of the PCE can be retained after 10 days. Graphene can also be coated onto n-type silicon nanowire arrays, where the nanowires suppress and harvest light much better than their planar counterparts. The interface of the heterojunctions is the most important part and a single layer of CVD-grown graphene with 97% transparency and sheet resistance of 350 Ωm^{-2} on silicon can exhibit a PCE of 5.38–7.85% which is much greater than multilayer graphene heterojunctions. This value can be increased even further to 8.94% by the incorporation of an antireflection layer of silicon dioxide. The thickness of these layers can tune the photovoltaic performance, with thicker layers producing a higher increase in the PCE. This layer also improves the stability of the solar cell under solar exposure, moisture and air. Graphene quantum dots and crystalline silicon can be used as electron blocking layers to prevent charge carrier recombination at solar cell anodes. In these cells, surface passivation can occur due to differing terminal groups, namely, oxide, hydrogen and methyl moieties. Graphene oxide can be used with gold nanoparticles to produce anodic buffer layers. Capping agents are utilized in these hybrids, generally in the form of glycine or sodium citrate and can show a PCE ranging from 2.82–3.34%. However, the inclusion of P3HT and IBCA into the solar cell can increase the PCE up to 5.10%. Graphene oxide nanoribbons (GORs) can be used a hole extraction layers in many solar cells. These layers have been developed to replace existing ITO-based materials and have so far managed to increase the PCE of a solar device from 2.20% to 4.19%. One of the most efficient graphene photoactive layers is produced from a hybrid material containing graphene oxide, PEDOT:PSS and n-type silicon nanowires [57].

The wt.% of graphene oxide has a profound effect on the PCE of the device with the optimum concentration being 30%, which produces a PCE of up to 9.57%. In comparison, the substitution of silicon nanowires for planar silicon produces a massive drop in the PCE to 4.30%. These layers not only show a high optical transparency compared to non-graphene photoactive layers of similar composition, but also exhibit a reduction in the exciton decay. The PCE can be tuned by altering the number of graphene layers and by introducing doping effects into the material. The doping of P3HT to act as an electron blocking layer between the silicon and graphene layers has produced a PCE of up to 10.30%, with a low carrier recombination at the anode. These have shown to be one of the most promising options as a photoactive layer material in heterojunction solar cells. There are many other notable photoactive layers produced from graphene derivatives/hybrid materials, including Graphene-CdS-based materials that can exhibit a PCE of up to 7.5%; GO-ethylenedinitrobenzoyl electron acceptor materials demonstrate PCE up to 2.41%; thiolated reduced graphene oxide (TrGO) materials show PCEs up to 4.20%; Graphene and transition metal oxide composites achieve PCEs up to 7.3%; and Graphene-C60-silver-based hybrid materials attain PCEs up to 7.2% [22, 65].

2.9.5 Graphene-Schottky Junction Solar Cells

GaAs solar cells have been one of the most widely studied types of heterojunction, namely Schottky junction, solar cells. PCE of GaAs/graphene solar cell is still lower than the commercially available GaAs thin film solar cells [24]. GaAs has a superior band gap to silicon, with a charge carrier mobility that is six times higher. Pillar array patterned silicon substrate with Graphene hetero junctions shows a PCE of up to 1.96%. The nitric acid doping, the cell can achieve a PCE of up to 3.55. A Schottky junction solar cell made from CdS nanowires and graphene has only achieved a PCE of 1.65%. By optimizing the open circuit voltage, junction ideality factor, graphene resistance, and internal interfacial contact, there is a possibility to achieve a PCE of up to 25.8% with these solar cells. Solar cells composed of Graphene/semiconductor van der Waals Schottky diodes, with a tuneable gate and Fermi level, lead the way in terms of efficiency [29]. Heterojunction utilizes a Graphene dielectric-Graphene gate to achieve a PCE of up to 18.5% much higher than other GaAs solar cells. The open circuit voltage, while not the best compared to other solar cell classes, is better than many other GaAs solar cells, with a value of 0.96 V. So, with a bit of optimization, and despite the discrepancies in quality over the whole class of solar cells, some GaAs could reach efficiencies comparable to commercial solar cells [21, 66].

3 Energy Storage Devices

The development of electronic components and its technology has been progressing day by day very rapidly over the last 20 years. Power storage technology solutions such as batteries and capacitors have been the primary resources and limiting factors due to size, power capacity and efficiency. Lithium-ion batteries face a trade-off between energy density and power density [7]. Graphene is a material that has commercial viability in small-scale energy storage systems, such as batteries and capacitors, which are used in many electronics. Graphene is a 2D material composed of all carbon atoms arranged in a hexagonal lattice with high electrical conductivity and charge carrier mobility, as well as high stability to temperature, chemicals, and other stimuli, so it enables interest across various energy storage devices [1].

Major energy storage devices are classified into three categories; they are capacitor/supercapacitors, batteries and fuel cells. Graphene has attracted great interest for supercapacitors because of its extraordinarily high surface area of up to 2,630 m^2g^{-1} [67]. Graphene-based supercapacitors have a maximum energy density which depends on several parameters. Those are thickness, density of the graphene film, current collector, separator, the nature and density of the electrolyte, the operating voltage window of the cell and the packaging efficiency. Graphene supercapacitors represent the next wave of energy storage technology, promising vastly superior performance to existing chemical batteries.

Graphene is used in a range of batteries including redox flow, metal–air, lithium–sulfur and lithium-ion batteries. It can be processed chemically into many forms which will be suitable for both the positive and negative electrodes, enabling the fabrication of an all-graphene battery with an ultrahigh energy density. Graphene composite materials, such as Carbon Nanotubes/Graphene sandwiches for high-rate lithium-sulfur batteries have been studied by many researchers. Graphene in the field of energy storage has an excellent projection. Its incorporation in the formulation of electrodes has led to tremendous improvements of specific properties in supercapacitors and batteries.

In supercapacitors, graphene represents an alternative to activated carbons with high specific surfaces. Its open, non-porous laminar structure allows the adsorption of many ions for the formation of the electric double layer quickly and reversibly. This has been combined with its high electronic conductivity, makes it possible to increase the power of the devices enabling their use in applications. Graphene can also contribute to the improvement of hybrid supercapacitors; devices, in terms of energy density, power density, and cyclability. The integration of Graphene in new innovative energy storage technologies such as lithium-sulfur or metal-air batteries improves some of their current limitations, such as cyclability or power, and will enable their development in the medium to long term. This strategy not only leads to obtaining electrodes with greater specific surface capacities and also the processing of electrodes does not need binding agents; thus reducing the weight of the final device and increasing the device´s gravimetric energy density. Currently, an enormous effort has been taken on next-generation chemical battery technology, including various forms of lithium ion batteries; several challenges have been prevented in the replacement of lithium ion batteries by supercapacitors.

3.1 Graphene-Based Batteries

Graphene-based batteries are among the most developed energy storage device. Graphene could dramatically increase the lifespan of a traditional lithium ion battery, meaning devices can be charged quicker and hold more power for longer. Batteries could be so flexible and light that they could be stitched into clothing or body for wearable electronics applications. Less weight and using batteries that can be recharged by body heat or the sun would allow them to stay out in the field for longer. In lithium-ion batteries, graphitic carbon is used as the anode material, where it has high crystallinity and the arrangement of graphene layers leads to LiC_6 being formed. This allows for a high transfer of electrons between lithium and carbon which is stored between two sheets, and a high energy capacity. The development of next-generation flexible electronics, such as soft, portable electronic products, roll-up displays, wearable devices, implantable biomedical devices, and conform- able health-monitoring electronic skin, requires power sources that are flexible. Nano-sized materials can be used to prepare electrodes through suitable routes toward flexible batteries. It is believed that the charge-discharge rate of a lithium ion battery (LIB) depends critically on the migration rate of lithium ions and electrons through the electrolyte and bulk electrodes into active electrode materials [68]. The strategy to increase ion and electron transport kinetics in batteries mainly focused on seeking new electrode materials and designing conductive electrode structures with high ion and electron transport rates or reducing the path length over which electrons and lithium ions have to move by using nanomaterial [69, 70]. Graphene Foam (GF) network as both a highly conductive pathway for electrons and ions and a 3D interconnected current collector, thin lightweight, and flexible $LiFePO_4$ (LFP)/GF and $Li_4Ti_5O_{12}$ (LTO)/GF were studied. The LTO/GF and LFP/GF electrodes showed initial discharge capacities of 170 mAh/g and 164 mAh/g at 0.2 C, respectively. It indicates that the LTO/GF can be a good match with the LFP/GF for assembling a full battery. Using the flexible LFP/GF cathode and LTO/GF anode, we then built a thin, lightweight, and flexible LTO/GF//LFP/GF full battery [71].

Graphene materials act as electrodes, on the one hand, can actively take part in electrochemical reactions. On the other hand, they can act as conductive additives to improve kinetics and as buffers to support the structural integrity of the electrodes as shown in Fig. 7 [72, 73]. The battery performance can be enhanced by doing alternate to the anode (graphite) with substances, such as Sn, Sb, Si, Ge, SnO_2, and Co_3O_4. But a major drawback of these substances as anode materials is the huge volume variation during the charge/discharge process which causes the pulverization of the electrode, resulting in poor reversibility. Metal oxide nanoparticles encapsulated by graphene layers have displayed high specific capacity and excellent cycling performance as anode materials in rechargeable lithium batteries (RLBs) compared to pure form of nanoparticles of the metal oxides. The Graphene layers acted as both a "buffer zone" of volume variation of the nanoparticles and a good electron transfer medium. The buffer layer around the electrode materials with functions of both excellent mechanical properties and good electron transportation behavior is of great potential for solving effectively the problems mentioned above [74]. From previous works, it has been found that high energy density and good electrochemical performance play vital roles in graphene-based electrodes for further applications in Sodium-ion batteries (SIBs). The output voltage of SIBs is directly related to the working potentials of anode and cathode materials. The operating potential of anode materials should be low but not very close to 0 V like examples of pure graphene, $Na_2Ti_3O_7$/RGO, $Sn4P_3$/RGO, and Na_2TP/RGO. For the cathode materials, the working potential should be high, like examples of $Na_3(VO_{0.5})_2(PO_4)_2F_2$/G and $NaVPO_4F$/G [49]. Similarly, the specific capacity must be as high as possible. The stable capacity of P/G composite is more than 1000 mA h g^{-1}, which is the highest value among all the reported graphene-based nanomaterial for SIBs [75].

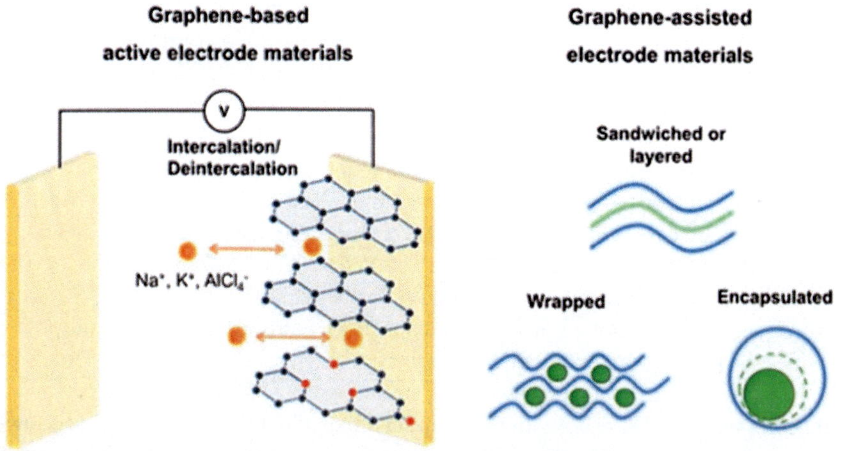

Fig. 7 Schematic diagram of the battery with graphene. Reproduced with permission from Comparing Graphene Battery-Australian Graphene Industry Association [73]

3.2 Graphene-Based Capacitors and Supercapacitors

Supercapacitors are also known as electric double-layer capacitors (EDLC) or Ultracapacitors. They differ from regular capacitors in that they can store large amounts of energy. Supercapacitors also contain two metal plates, coated with a porous material like activated carbon. They are immersed in an electrolyte made up of positive and negative ions dissolved in a solvent. The electrodes which, one plate is positive and the other is negative are connected to the circuit. During charging, ions from the electrolyte accumulate on the surface of each carbon coated electrode plate. Supercapacitors stores energy in an electric field that is formed between two oppositely charged particles, only they have the electrolyte. Thus, during charging each electrode has two layers of charge coating, it is an electric double layer. Supercapacitors boast a high energy storage capacity compared to regular capacitors, but they still lag behind batteries in that area. Graphene-based supercapacitors could provide massive amounts of power while using much less energy than conventional devices. Graphene can be used as an alternative to the present materials which are storing ions on supercapacitor electrodes because of its high surface area, stability, and conductivity. Graphene could be used as a coating on an electrode plate to form a double layer coating, or as an electrode to improve the electrode surface area by Graphene 2D material and enhance the network conductivity, leading to better ionic transport via the electrolyte. It will be possible one day; supercapacitors could offer a viable alternative to Li-ion batteries. Graphene supercapacitor is said to store almost as much energy as a lithium-ion battery, charge and discharge in seconds, and maintain all this over tens of thousands of charging cycles. One of the ways to achieve this is by using a highly porous form of graphene with a large internal surface area which can be made by packing graphene powder into a coin shaped cell.

There are many nanocomposites with graphene and its associated materials integrated by the metal oxide nanoparticles. They offer the hybrid form of nanocomposites which are showing very high performance such as 1D $NiCo_2O_4$ nanowire coated reduced graphene Oxide (rGO) can deliver a high specific capacitance of 1248 Fg^{-1} due to its unique interfacial effect with electrolytes [76]. Apart from the diversity of metal oxides, the construction of metal oxides and other materials, such as hydroxides, metal quantum dots and organics, has also emerged over the past years. The chemical composition of metal oxide-based electrodes varies diversely, but fully understanding the detailed synergistic effect from different components to rationally design the electrodes still deserves further study [77].

RuO_2 shows great potential for supercapacitors with higher energy and power densities than carbon-based EDLCs and polymer-based pseudocapacitors. The exfoliated graphene nanosheets as the substrate to hydrous ruthenium oxide (RuO_2) yield composites (ROGSCs) with different loadings of Ru through sol–gel and low-temperature annealing processes. ROGSCs with 38.3 wt.% of Ru have a 281 m^2/g specific surface area than that of pure graphene sheets of 108 m^2/g. ROGSCs with 38.3 wt.% of Ru exhibited a high specific capacitance of ~570 F/g at 1 mV/s, enhanced rate capability, and high energy density of 20.1 W h/kg at low operation rate of 100 mA/g or high power density of 10 000 W/kg at a reasonable energy density of 4.3 W h/kg [78]. The electrodes that are based on $Ni(OH)_2$–Graphene composites showed a specific capacitance of ~1335 F/g at a charge and discharge current density of 2.8 A/g [79]. Fe_3O_4/(Graphene oxide) nanocomposites with 73.5% Fe_3O_4 nanoparticles displayed a 480 F/g at a discharge current density of 5 A/g with the corresponding energy density of 67 W h/kg at a power density of 5506 W/kg and 843 F/g at a discharge current density of 1 A/g with the corresponding energy density of 124 W h/kg at a power density of 332 W/kg [80]. Graphene–$Co(OH)_2$ nanocomposites in a water–isopropanol system achieved the high electrochemical specific capacitance of 972.5 F/g at the current density of 500 mA/g. The decoration of graphene sheets with $Co(OH)_2$ nanocrystals enhances the electrochemical performance of the nanocomposites effectively due to the relatively higher utilization of $Co(OH)_2$ [50]. Finally, it has been observed that the improvement of the electrochemical performance of the graphene-based composite electrodes happened due to graphene in composites can support the active components, next, it provides the electronic conductive channels, and excellent interfacial contact between the active component and finally unique structural and mechanical properties of graphene restrict the mechanical deformation of the active component during the redox process as shown in Fig. 7.

3.3 *Hybridization of Battery-Electrochemical Capacitors*

The most intuitive approach to combine high energy and high power density within a single device is to combine the different types of energy storage sources. Attaching EDLCs and LIBs may provide the battery capacitor hybrids. This is not as common as the two areas mentioned earlier, this has emerged which combines both batteries and capacitors into a single hybrid device. As mentioned earlier, one of the reasons why supercapacitors have not been widely used compared to conventional capacitors and other energy storage mediums is down to cost. One way of reducing the cost has been to create hybrid storage devices which utilize the strength of Li-ion batteries with the rapid charging ability of supercapacitors. This can be achieved by integrating Graphene-based supercapacitors into Li-ion modules to increase the energy density, charge, and discharge cycle rates, and stability against the appropriate individual constituents [81]. As an example to materialize this idea, the hybridized lithium iron phosphate ($LiFePO_4$) battery material with poly(2,2,6,6-tetramethyl-1-piperinidyloxy-4-ylmethacrylate) (PTMA) redox capacitor has taken here. The hybrid electrode displays two distinct charge–discharge which is consistent with redox processes in $LiFePO_4$ and PTMA constituents. This hybrid electrode shows dramatic cycling stability enhancement at high recharge-discharge rates: only 12% capacity loss after 1,500 cycles at a rate of 5 C. The electrochemical potential of the PTMA/PTMA + redox couple is 3.63 V *versus* Li/Li^+ which is higher than that of $LiFePO_4/FePO_4$ of 3.44 V *versus* Li/Li^+ [82]. This state's hybridization of the two separate components yields a remarkable set of properties.

4 Battery-Powered Vehicles

There are four types of electric vehicles available. They are Battery Electric Vehicle (BEV) which fully powered by electricity. These are more efficient compared to hybrid and plug-in hybrids. Hybrid Electric Vehicle (HEV): The vehicle uses both the internal combustion (usually petrol) engine and the battery-powered motor power train. The petrol engine is used both to drive and charge when the battery is empty. These vehicles are not as efficient as fully electric or plug-in hybrid vehicles. Plug-in Hybrid Electric Vehicle (PHEV): Uses both an internal combustion engine and a battery charged from an external socket (they have a plug). This means the vehicle's battery can be charged with electricity rather than the engine. PHEVs are more efficient than HEVs but less efficient than BEVs. Fuel Cell Electric Vehicle (FCEV): Electric energy is produced from chemical energy [83]. Electric Vehicle (EVs) will have a very important role in smart cities, along with shared mobility, public transport, etc. Therefore, more efforts have been taken by the scientists to facilitate the charging process and to improve batteries. Energy storage systems of batteries play a critical role in EVs, plug-in hybrid electric vehicles (PHEVs), and hybrid electric vehicles (HEVs). In EVs, batteries act as the only energy source where electrical energy is stored. The major types of batteries currently used in EVs, PHEVs, and HEVs as energy storage systems include lithium-ion batteries, nickel-metal hydride batteries, lead-acid batteries, and ultracapacitors [84]. One of the major assets is graphene-based batteries that barely heat, enabling fast or ultra-fast charges without significant power losses due to heat. In a high power plug, this battery could be charged in only 5 min. This kind of battery is in development, they exist as prototypes of graphene batteries with a specific power of 1 kWh/kg, but it is to be expected about 6.4 kWh/kg [85]. Our previous discussion made detailed discussions about the role of graphene in batteries, supercapacitors and its hybrids. The developments done on the batteries and supercapacitors will impact the development of EVs (Fig. 8).

5 Fuel-Cell Technology

A fuel cell is an electrochemical cell that converts the chemical energy of a fuel often hydrogen and an oxidizing agent often oxygen into electricity through a pair of redox reactions. Fuel cells can produce electricity continuously for as long as fuel and oxygen are supplied. In order to make hydrogen fuel cells a viable energy source for the future, material scientists finds a way to integrate this technology with graphene. Hydrogen fuel cells are converting chemical energy to electrical energy with only one by-product of water. To minimize the cost of fuel cells, carbon allotropes can be used as supports for the platinum catalysts. Graphene is one such catalyst support. Graphene oxide provides good dispersion, large surface area and high conductivity when it is reduced. So the integrity and efficiency of the device is enhanced by integrating Graphene into fuel cells. It helps to facilitate the oxidation of methanol and

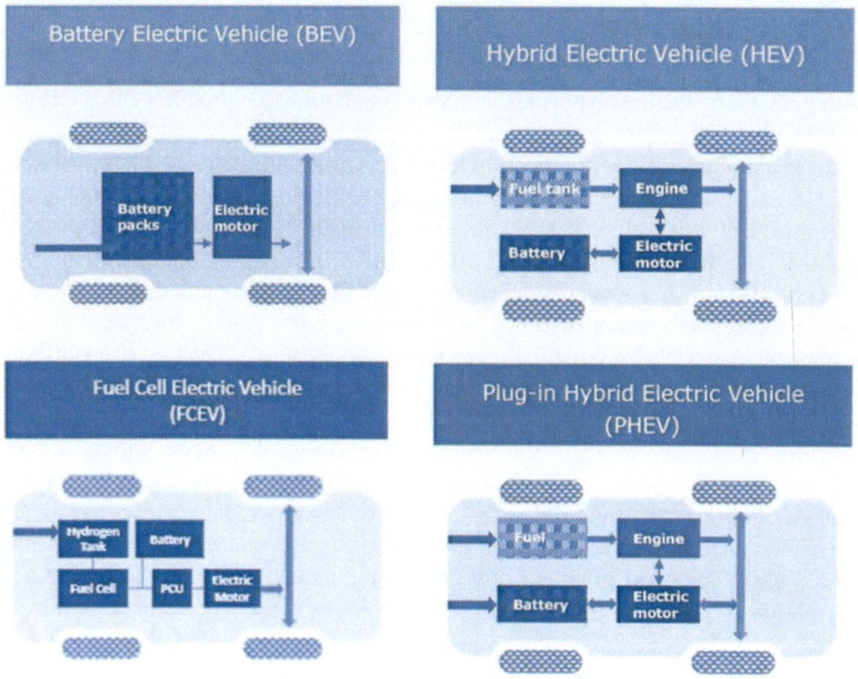

Fig. 8 Schematic diagrams of the types of electric vehicles [83] (https://e-amrit.niti.gov.in/types-of-electric-vehicles)

has been found to be much more effective than other carbon allotropes, such as Carbon Nanotubes and carbon black. Graphene's two-dimensional sheets provide a greater active surface area for electron/ion transport, as both sides of the sheet are exposed to the solution within the fuel cell [86]. The high surface area and high conductive properties of graphene-based nanomaterial are promising to use as electrocatalysts for oxygen reduction reaction (ORR) and fuel oxidation. The polymer membranes combined with graphene possess high ionic conductivity, high tensile strength, and low fuel permeation [87]. For example, a 22 μm-thick ozonated GO film was obtained by bubbling O_3 gas of 5 mg mL^{-1} GO solution, followed by vacuum filtration through a 0.22 mm cellulose membrane. The layer-by-layer assembling was conducted to fabricate graphene on the membrane electrode assembly (MEA). Zhang et al. [88] reported that atomically dispersed Ru on N-doped graphene exhibited high ORR activity, better durability, and tolerance toward methanol and CO poisoning than commercial Pt/C catalysts in 0.1 mol/L $HClO_4$. In another work, various atomically dispersed metals were immobilized in N-doped graphene lattices by a facile annealing strategy for ORR, such as Co, Fe, Cu, Mn, Sc, Rh, and Ru. In it, a certain amount of metal salts and GO were well dispersed in water and then water was removed by rotary evaporator which shows the better performance [87].

6 Other Applications

Many scientific evidences have indicated that graphene-based products can improve the efficiency of green energy technologies and energy storage systems that are essential to the use of intermittent energy sources such as solar and wind. Lot of efforts have been dedicated to graphene-based green energy technology. The new technologies for graphene production, such as 3D printing, may lead to significant breakthroughs. 3D printing termed as additive manufacturing is the "process of joining materials to make objects from 3D model data, usually layer upon layer, as opposed to subtractive manufacturing methodologies [89]. The following are few recent other areas of use of graphene materials for potential green energy applications. The first one is using bacteria to create graphene nanomaterial solar technology, a novel biomineralization method for creating essential metal sulfide nanoparticles. Next is, a reduced graphene oxide supported photocatalyst that cleaves water to form hydrogen, which can be used as a fuel, for climate change applications. Also, algae biomass is emerging as an exciting area of study due to its ability to convert solar radiation into high energy biomolecules using GO based materials [90]. Beyond this, there are many other applications using graphene, and will be a potential candidate for the future technology.

7 Limitations and Future Perspectives

The production of graphene is bit difficult and expensive; hence the production cost of the technology will be higher. It is the much considerable limitation of graphene. Otherwise, it has numerous advantages rather than the limitations. Additionally, the reactive nature of graphene with oxygen and heat deforms to the oxidized state from the pure form of graphene which reduces the conducting properties. Further, the thickness of the graphene can be controlled by the few layers of graphene for various applications which can be researched further.

8 Conclusions

The graphene is most important conducting material in the 2D form for various applications for energy production and energy storage applications. This book chapter discussed about the recent importance of graphene for applications. It can have application on supercapacitors, fuel cells, batteries, and various types of solar photovoltaic cells. Pure graphene can be converted into a semiconducting material by adding suitable dopants to change its conducting properties. Finally, other areas of applications like 3D printing, biological applications, and the limitations. It has been understood that graphene will be unavoidable 2D layered material for energy-related technologies in the future.

References

1. Tiwari, S.K., Sahoo, S., Wang, N., Huczko, A.: Graphene research and their outputs: status and prospect. J. Sci. Adv. Mater. Devices **5**, 10–29 (2020). https://doi.org/10.1016/J.JSAMD.2020.01.006
2. Sang, M., Shin, J., Kim, K., Yu, K.J.: Electronic and thermal properties of graphene and recent advances in graphene based electronics applications. Nanomaterials **9**, (2019). https://doi.org/10.3390/NANO9030374
3. Yang, G., Zhang, Y., Yan, X., Yang, G., Zhang, Y., Yan, X.: Electronic structure and optical properties of a new type of semiconductor material: graphene monoxide. J. Semicond. 34(8), 083004–083005 (2013). https://doi.org/10.1088/1674-4926/34/8/083004
4. Zhang, Y., Tang, T.T., Girit, C., Hao, Z., Martin, M.C., Zettl, A., Crommie, M.F., Shen, Y.R., Wang, F.: Direct observation of a widely tunable bandgap in bilayer graphene. Nature **459**, 820–823 (2009). https://doi.org/10.1038/NATURE08105
5. Kim, M., Xu, S.G., Berdyugin, A.I., Principi, A., Slizovskiy, S., Xin, N., Kumaravadivel, P., Kuang, W., Hamer, M., Krishna Kumar, R., Gorbachev, R.V., Watanabe, K., Taniguchi, T., Grigorieva, I.V., Fal, V.I., Polini, M., Geim, A.K.: Control of electron-electron interaction in graphene by proximity screening. https://doi.org/10.1038/s41467-020-15829-1
6. Guler, O., Bagci, N.: A short review on mechanical properties of graphene reinforced metal matrix composites. J. Mater. Res. Technol. **9**, 6808–6833 (2020). https://doi.org/10.1016/J.JMRT.2020.01.077
7. Hossain, S., Abdalla, A.M., Suhaili, S.B.H., Kamal, I., Shaikh, S.P.S., Dawood, M.K., Azad, A.K.: Nanostructured graphene materials utilization in fuel cells and batteries: a review. J. Energy Storage **29**, (2020). https://doi.org/10.1016/j.est.2020.101386
8. Czerniak-Reczulska, M., Niedzielska, A., Jędrzejczak, A.: Graphene as a material for solar cells applications. Adv. Mater. Sci. **15**, 67–81 (2015). https://doi.org/10.1515/ADMS-2015-0024
9. Tang, Q., Wang, X., Yang, P., He, B.: A solar cell that is triggered by sun and rain. Angew. Chemie. **128**, 5329–5332 (2016). https://doi.org/10.1002/ANGE.201602114
10. Graphene solar panels work rain or shine. Engineers Australia, https://www.engineersaustralia.org.au/News/graphene-solar-panels-work-rain-or-shine
11. Shi, E., Li, H., Yang, L., Zhang, L., Li, Z., Li, P., Shang, Y., Wu, S., Li, X., Wei, J., Wang, K., Zhu, H., Wu, D., Fang, Y., Cao, A.: Colloidal antireflection coating improves graphene-silicon solar cells. Nano Lett. **13**, 1776–1781 (2013). https://doi.org/10.1021/NL400353F/SUPPL_FILE/NL400353F_SI_001.PDF
12. Arora, N., Dar, M.I., Akin, S., Uchida, R., Baumeler, T., Liu, Y., Zakeeruddin, S.M., Grätzel, M.: Low-cost and highly efficient carbon-based perovskite solar cells exhibiting excellent long-term operational and UV stability. Small **15**, 1904746 (2019). https://doi.org/10.1002/SMLL.201904746
13. Kim, S.B., Park, J.Y., Kim, C.S., Okuyama, K., Lee, S.E., Jang, H.D., Kim, T.O.: Effects of graphene in dye-sensitized solar cells based on nitrogen-doped TiO_2 composite. J. Phys. Chem. C **119**, 16552–16559 (2015). https://doi.org/10.1021/ACS.JPCC.5B02309/SUPPL_FILE/JP5B02309_SI_001.PDF
14. Won, R.: Graphene–silicon solar cells. Nat. Photonics **47**(4), 411–411 (2010). https://doi.org/10.1038/nphoton.2010.140
15. Lin, X.F., Zhang, Z.Y., Yuan, Z.K., Li, J., Xiao, X.F., Hong, W., Chen, X.D., Yu, D.S.: Graphene-based materials for polymer solar cells. Chin. Chem. Lett. **27**, 1259–1270 (2016). https://doi.org/10.1016/J.CCLET.2016.06.041

16. Muchuweni, E., Martincigh, B.S., Nyamori, V.O.: Recent advances in graphene-based materials for dye-sensitized solar cell fabrication. RSC Adv. **10**, 44453–44469 (2020). https://doi.org/10.1039/D0RA08851J
17. Wu, W., Wu, H., Zhong, M., Guo, S.: Dual role of graphene quantum dots in active layer of inverted bulk heterojunction organic photovoltaic devices. ACS Omega **4**, 16159–16165 (2019). https://doi.org/10.1021/ACSOMEGA.9B02348/SUPPL_FILE/AO9B02348_SI_001.PDF
18. Lamanna, E., Matteocci, F., Calabrò, E., Serenelli, L., Salza, E., Martini, L., Menchini, F., Izzi, M., Agresti, A., Pescetelli, S., Bellani, S., Del Río Castillo, A.E., Bonaccorso, F., Tucci, M., Di Carlo, A.: Mechanically stacked, two-terminal graphene-based perovskite/silicon tandem solar cell with efficiency over 26%. Joule. **4**, 865–881 (2020). https://doi.org/10.1016/J.JOULE.2020.01.015
19. Gong, K., Hu, J., Cui, N., Xue, Y., Li, L., Long, G., Lin, S.: The roles of graphene and its derivatives in perovskite solar cells: a review. Mater. Des. **211**, 110170 (2021). https://doi.org/10.1016/J.MATDES.2021.110170
20. Amollo, T.A., Mola, G.T., Nyamori, V.O.: Organic solar cells: materials and prospects of graphene for active and interfacial layers. Crit. Rev. Solid State Mater. Sci. **45**, 261–288 (2019). https://doi.org/10.1080/10408436.2019.1632791
21. Singh, E., Nalwa, H.S.: Stability of graphene-based heterojunction solar cells. RSC Adv. **5**, 73575–73600 (2015). https://doi.org/10.1039/C5RA11771B
22. Li, P., Chen, C., Zhang, J., Li, S., Sun, B., Bao, Q.: Graphene-based transparent electrodes for hybrid solar cells. Front. Mater. **1**, 26 (2014). https://doi.org/10.3389/FMATS.2014.00026/BIBTEX
23. Wang, H.X., Wang, Q., Zhou, K.G., Zhang, H.L.: Graphene in light: design, synthesis and applications of photo-active graphene and graphene-like materials. Small **9**, 1266–1283 (2013). https://doi.org/10.1002/SMLL.201203040
24. Wen, L., Gao, F., Yu, Y., Xu, Z., Liu, Z., Gao, P., Zhang, S., Li, G.: Enhancing the photovoltaic performance of GaAs/graphene Schottky junction solar cells by interfacial modification with self assembled alkyl thiol monolayer. J. Mater. Chem. A **6**, 17361–17370 (2018). https://doi.org/10.1039/C8TA04490B
25. Zhang, C., Wang, S., Zhang, H., Feng, Y., Tian, W., Yan, Y., Bian, J., Wang, Y., Jin, S., Zakeeruddin, S.M., Grätzel, M., Shi, Y.: Efficient stable graphene-based perovskite solar cells with high flexibility in device assembling via modular architecture design. Energy Environ. Sci. **12**, 3585–3594 (2019). https://doi.org/10.1039/C9EE02391G
26. You, P., Liu, Z., Tai, Q., Liu, S., Yan, F.: Efficient semitransparent perovskite solar cells with graphene electrodes. Adv. Mater. **27**, 3632–3638 (2015). https://doi.org/10.1002/ADMA.201501145
27. Hadadian, M., Correa-Baena, J.P., Goharshadi, E.K., Ummadisingu, A., Seo, J.Y., Luo, J., Gholipour, S., Zakeeruddin, S.M., Saliba, M., Abate, A., Grätzel, M., Hagfeldt, A.: Enhancing efficiency of perovskite solar cells via N-doped graphene: crystal modification and surface passivation. Adv. Mater. **28**, 8681–8686 (2016). https://doi.org/10.1002/ADMA.201602785
28. Iqbal, M.Z., Rehman, A.U.: Recent progress in graphene incorporated solar cell devices. Sol. Energy **169**, 634–647 (2018). https://doi.org/10.1016/J.SOLENER.2018.04.041
29. Rehman, M.A., Roy, S.B., Akhtar, I., Bhopal, M.F., Choi, W., Nazir, G., Khan, M.F., Kumar, S., Eom, J., Chun, S.H., Seo, Y.: Thickness-dependent efficiency of directly grown graphene based solar cells. Carbon N. Y. **148**, 187–195 (2019). https://doi.org/10.1016/J.CARBON.2019.03.079
30. Bointon, T.H., Russo, S., Craciun, M.F.: Is graphene a good transparent electrode for photovoltaics and display applications? https://doi.org/10.1049/iet-cds.2015.0121
31. Ramakrishna Matte, H.S.S., Subrahmanyam, K.S., Rao, C.N.R.: Synthetic aspects and selected properties of graphene invited feature article
32. Shastry, T.A., Balla, I., Bergeron, H., Amsterdam, S.H., Marks, T.J., Hersam, M.C.: Mutual photoluminescence quenching and photovoltaic effect in large-area single-layer MoS_2−polymer heterojunctions article (2016). https://doi.org/10.1021/acsnano.6b06592

33. Wang, X., Sun, G., Routh, P., Kim, D.H., Huang, W., Chen, P.: Heteroatom-doped graphene materials: syntheses, properties and applications. Chem. Soc. Rev. **43**, 7067–7098 (2014). https://doi.org/10.1039/C4CS00141A
34. Chae, S., Panomsuwan, G., Bratescu, M.A., Teshima, K., Saito, N.: p-Type doping of graphene with cationic nitrogen. ACS Appl. Nano Mater. **25**, 37 (2022). https://doi.org/10.1021/acsanm.8b02237
35. Priyadarsini, A., Mallik, B.S.: Effects of doped N, B, P, and S atoms on graphene toward oxygen evolution. Reactions **6**, 5368–5378 (2021). https://doi.org/10.1021/acsomega.0c05538
36. Kong, X., Zhang, L., Liu, B., Gao, H., Zhang, Y., Yan, H., Song, X.: Graphene/Si Schottky solar cells: a review of recent advances and prospects. RSC Adv. **9**, 863–877 (2019). https://doi.org/10.1039/C8RA08035F
37. Li, C., Ma, Y., Zhang, X., Chen, X., Xi, F., Li, S., Ma, W., Chang, Y.: Enhanced efficiency of graphene-silicon schottky junction solar cell through pyramid arrays texturation. Silicon **1**, 3. https://doi.org/10.1007/s12633-021-01579-2
38. Wei, D., Kivioja, J.: Graphene for energy solutions and its industrialization. Nanoscale **5**, 10108–10126 (2013). https://doi.org/10.1039/C3NR03312K
39. Koh, W.S., Gan, C.H., Phua, W.K., Akimov, Y.A., Bai, P.: The potential of graphene as an ITO replacement in organic solar cells: an optical perspective. IEEE J. Sel. Top. Quantum Electron. **20**, (2014). https://doi.org/10.1109/JSTQE.2013.2247976
40. Behura, S.K., Wang, C., Wen, Y., Berry, V.: Graphene–semiconductor heterojunction sheds light on emerging photovoltaics. Nat. Photonics **135**(13), 312–318 (2019). https://doi.org/10.1038/s41566-019-0391-9
41. Bramhaiah, K., Bhattacharyya, S.: Challenges and future prospects of graphene-based hybrids for solar fuel generation: moving towards next generation photocatalysts. Mater. Adv. **3**, 142 (2022). https://doi.org/10.1039/d1ma00748c
42. Pang, S., Hernandez, Y., Feng, X., Müllen, K.: Graphene as transparent electrode material for organic electronics. Adv. Mater. **23**, 2779–2795 (2011). https://doi.org/10.1002/ADMA.201100304
43. Patel, K., Tyagi, P.K.: Multilayer graphene as a transparent conducting electrode in silicon heterojunction solar cells. AIP Adv. **5**, 077165 (2015). https://doi.org/10.1063/1.4927545
44. McConnell, R.D.: Assessment of the dye-sensitized solar cell. Renew. Sustain. Energy Rev. **6**, 271–293 (2002). https://doi.org/10.1016/S1364-0321(01)00012-0
45. Sharma, K., Sharma, V., Sharma, S.S.: Dye-sensitized solar cells: fundamentals and current status (2018). https://link.springer.com/article/10.1186/s11671-018-2760-6
46. Guo, X., Lu, G., Chen, J.: Graphene-based materials for photoanodes in dye-sensitized solar cells (2015)
47. Dhonde, M., Sahu, K., Murty, V.V.S.: Cu-doped TiO_2 nanoparticles/graphene composites for efficient dye-sensitized solar cells. Sol. Energy **220**, 418–424 (2021). https://doi.org/10.1016/J.SOLENER.2021.03.072
48. Lin, K.-Y., Nguyen, M.T., Waki, K., Jiang, J.-C.: Boron and nitrogen co-doped graphene used as counter electrode for iodine reduction in dye-sensitized solar cells (2018). https://doi.org/10.1021/acs.jpcc.8b06956
49. Sun, Y., Wang, S., Li, C., Luo, P., Tao, L., Wei, Y., Shi, G.: Large scale preparation of graphene quantum dots from graphite with tunable fluorescence properties. Phys. Chem. Chem. Phys. **15**, 9907–9913 (2013). https://doi.org/10.1039/c3cp50691f
50. Chen, S., Zhu, J., Wang, X.: One-step synthesis of graphene-cobalt hydroxide nanocomposites and their electrochemical properties. https://doi.org/10.1021/jp1048474
51. Jo, I.R., Lee, Y.H., Kim, H., Ahn, K.S.: Multifunctional nitrogen-doped graphene quantum dots incorporated into mesoporous TiO_2 films for quantum dot-sensitized solar cells. J. Alloys Compd. **870**, 159527 (2021). https://doi.org/10.1016/J.JALLCOM.2021.159527
52. Xue, J., Long, L., Zhang, L., Luo, H., Yang, L., Liu, F., Lv, F., Kong, W., Liu, J.: Enhanced H_2 evolution and the interfacial electron transfer mechanism of titanate nanotube sensitized with CdS quantum dots and graphene quantum dots. Int. J. Hydrog. Energy **45**, 6476–6486 (2020). https://doi.org/10.1016/J.IJHYDENE.2019.12.196

53. Redondo-Obispo, C., Serafini, P., Climent-Pascual, E., Ripolles, T.S., Mora-Seró, I., De Andrés, A., Coya, C.: Effect of pristine graphene on methylammonium lead iodide films and implications on solar cell performance (2021). https://doi.org/10.1021/acsaem.1c02738
54. Mariani, P., Najafi, L., Bianca, G., Zappia, I., Gabatel, L., Agresti, A., Pescetelli, S., Carlo, A.D., Bellani, S., Bonaccorso, F.: Low-temperature graphene-based paste for large-area carbon perovskite solar cells. ACS Appl. Mater. Interfaces. **13**, 33 (2021). https://doi.org/10.1021/acsami.1c02626
55. O'keeffe, P., Catone, D., Paladini, A., Toschi, F., Turchini, S., Avaldi, L., Martelli, F., Agresti, A., Pescetelli, S., Del, A.E., Castillo, R., Bonaccorso, F., Di Carlo, A.: Graphene-induced improvements of perovskite solar cell stability: effects on hot-carriers (2019). https://doi.org/10.1021/acs.nanolett.8b03685
56. Yamaguchi, M., Lee, K.H., Araki, K., Kojima, N.: A review of recent progress in heterogeneous silicon tandem solar cells. J. Phys. D Appl. Phys. **51**, 133002 (2018). https://doi.org/10.1088/1361-6463/AAAF08
57. Anrango-Camacho, C., Pavón-Ipiales, K., Frontana-Uribe, B.A., Palma-Cando, A.: Recent advances in hole-transporting layers for organic solar cells. Nanomaterials **12**, 443 (2022). https://doi.org/10.3390/NANO12030443
58. Yang, Q.D., Ng, T.W., Lo, M.F., Wong, N.B., Lee, C.S.: Enhanced storage/operation stability of small molecule organic photovoltaics using graphene oxide interfacial layer. Org. Electron. **13**, 3220–3225 (2012). https://doi.org/10.1016/J.ORGEL.2012.09.034
59. Facchetti, A.: Polymer donor–polymer acceptor (all-polymer) solar cells. Mater. Today **16**, 123–132 (2013). https://doi.org/10.1016/J.MATTOD.2013.04.005
60. Hsiao, Y.S., Whang, W.T., Chen, C.P., Chen, Y.C.: High-conductivity poly(3,4-ethylenedioxythiophene): poly(styrene sulfonate) film for use in ITO-free polymer solar cells. J. Mater. Chem. **18**, 5948–5955 (2008). https://doi.org/10.1039/B813079E
61. Jiang, Y., Li, C., Liu, H., Qin, R., Ma, H.: Poly(3,4-ethylenedioxythiophene): poly(styrenesulfonate)(PEDOT:PSS)–molybdenum oxide composite films as hole conductors for efficient planar perovskite solar cells. J. Mater. Chem. A **4**, 9958–9966 (2016). https://doi.org/10.1039/C6TA03658A
62. Zhang, Y., Chen, L., Hu, X., Zhang, L., Chen, Y.: Low work-function poly(3,4-ethylenedioxylenethiophene): poly(styrene sulfonate) as electron-transport layer for high-efficient and stable polymer solar cells open. Sci. Rep. **5**, 12839 (2015). https://doi.org/10.1038/srep12839
63. Mustonen, P., Mackenzie, D.M.A., Lipsanen, H.: Review of fabrication methods of large-area transparent graphene electrodes for industry. Front. Optoelectron. **132**(13), 91–113 (2020). https://doi.org/10.1007/S12200-020-1011-5
64. Davoise, L.V., Díez-Pascual, A.M., Capilla, R.P.: Application of graphene-related materials in organic solar cells. Materials **15**, 1171 (2022). https://doi.org/10.3390/MA15031171
65. Li, X., Wang, Y., Wang, Y., Wang, H., Qi, X., He, J., Xiao, S.: Donor-acceptor type reduced graphene-oxide and a tin-selenide nanohybrid with broad and ultrafast optical limiting properties. Front. Phys. **8**, 298 (2020). https://doi.org/10.3389/FPHY.2020.00298/BIBTEX
66. He, X., Li, R., Liu, J., Liu, Q., Chen, R.R., Song, D., Wang, J.: Hierarchical $FeCo_2O_4$@NiCo layered double hydroxide core/shell nanowires for high performance flexible all-solid-state asymmetric supercapacitors. Chem. Eng. J. **334**, 1573–1583 (2018). https://doi.org/10.1016/J.CEJ.2017.11.089
67. Li, X., Zhi, L.: Graphene hybridization for energy storage applications. Chem. Soc. Rev. **47**, 3189–3216 (2018). https://doi.org/10.1039/C7CS00871F
68. Feckl, J.M., Fominykh, K., Döblinger, M., Fattakhova-Rohlfing, D., Bein, T.: Nanoscale porous framework of lithium titanate for ultrafast lithium insertion. Angew. Chem. Int. Ed. Engl. **51**, 7459–7463 (2012). https://doi.org/10.1002/ANIE.201201463
69. Aricò, A.S., Bruce, P., Scrosati, B., Tarascon, J.M., Van Schalkwijk, W.: Nanostructured materials for advanced energy conversion and storage devices. Nat. Mater. **45**(4), 366–377 (2005). https://doi.org/10.1038/nmat1368

70. Lee, S., Cho, Y., Song, H.K., Lee, K.T., Cho, J.: Carbon-coated single-crystal $LiMn_2O_4$ nanoparticle clusters as cathode material for high-energy and high-power lithium-ion batteries. Angew. Chemie Int. Ed. **51**, 8748–8752 (2012). https://doi.org/10.1002/ANIE.201203581
71. Li, N., Chen, Z., Ren, W., Li, F., Cheng, H.M.: Flexible graphene-based lithium ion batteries with ultrafast charge and discharge rates. Proc. Natl. Acad. Sci. **109**, 17360–17365 (2012). https://doi.org/10.1073/pnas.1210072109
72. Xu, H., Chen, H., Gao, C.: Advanced graphene materials for sodium/potassium/aluminum-ion batteries. ACS Mater. Lett. **3**, 1221–1237 (2021). https://doi.org/10.1021/ACSMATERIALSLETT.1C00280/ASSET/IMAGES/MEDIUM/TZ1C00280_0009.GIF
73. Comparing Graphene Battery-Australian Graphene Industry Association. https://grapheneindustry.org.au/2019/08/comparing-graphene-battery-with-lithium-ion-battery/
74. Liang, M., Zhi, L.: Graphene -based electrode materials for rechargeable lithium batteries. J. Mater. Chem. **19**, 5871–5878 (2009). https://doi.org/10.1039/B901551E
75. Lu, Y., Lu, Y., Niu, Z., Chen, J.: Graphene-based nanomaterials for sodium-ion batteries. Adv. Energy Mater. **8**, 1702469 (2018). https://doi.org/10.1002/AENM.201702469
76. An, C., Zhang, Y., Guo, H., Wang, Y.: Metal oxide-based supercapacitors: progress and prospectives. Nanoscale Adv. **1**, 4644–4658 (2019). https://doi.org/10.1039/C9NA00543A
77. Ding, J., Hu, W., Paek, E., Mitlin, D.: Review of hybrid ion capacitors: from aqueous to lithium to sodium (2018). https://doi.org/10.1021/acs.chemrev.8b00116
78. Huang, Y., Liang, J., Chen, Y.: An overview of the applications of graphene-based materials in supercapacitors. Small **8**, 1805–1834 (2012). https://doi.org/10.1002/smll.201102635
79. Wang, H., Casalongue, H.S., Liang, Y., Dai, H.: $Ni(OH)_2$ nanoplates grown on graphene as advanced electrochemical pseudocapacitor materials. J. Am. Chem. Soc. **132**, 7472–7477 (2010). https://doi.org/10.1021/JA102267J
80. Shi, W., Zhu, J., Sim, D.H., Tay, Y.Y., Lu, Z., Zhang, X., Sharma, Y., Srinivasan, M., Zhang, H., Hng, H.H., Yan, Q.: Achieving high specific charge capacitances in Fe_3O_4/reduced graphene oxide nanocomposites. https://doi.org/10.1039/c0jm03175e
81. Cericola, D., Kötz, R.: Hybridization of rechargeable batteries and electrochemical capacitors: principles and limits. Electrochim. Acta. **72**, 1–17 (2012). https://doi.org/10.1016/J.ELECTACTA.2012.03.151
82. Vlad, A., Singh, N., Rolland, J., Melinte, S., Ajayan, P.M., Gohy, J.-F.: Hybrid supercapacitor-battery materials for fast electrochemical charge storage (2014). https://doi.org/10.1038/srep04315
83. Types of electric vehicles. https://e-amrit.niti.gov.in/types-of-electric-vehicles
84. Sanguesa, J.A., Torres-Sanz, V., Garrido, P., Martinez, F.J., Marquez-Barja, J.M.: A review on electric vehicles: Technologies and challenges. Smart Cities. **4**, 372–404 (2021). https://doi.org/10.3390/smartcities4010022
85. Kim, H., Park, K.-Y., Hong, J., Kang, K.: All-graphene-battery: bridging the gap between supercapacitors and lithium ion batteries (2014). https://doi.org/10.1038/srep05278
86. Iqbal, M.Z., Rehman, A.U., Siddique, S.: Prospects and challenges of graphene based fuel cells. J. Energy Chem. **39**, 217–234 (2019). https://doi.org/10.1016/j.jechem.2019.02.009
87. Su, H., Hu, Y.H.: Recent advances in graphene-based materials for fuel cell applications. Energy Sci. Eng. **9**, 958–983 (2021). https://doi.org/10.1002/ESE3.833
88. Zhang, C., Sha, J., Fei, H., Liu, M., Yazdi, S., Zhang, J., Zhong, Q., Zou, X., Zhao, N., Yu, H., Jiang, Z., Ringe, E., Yakobson, B.I., Dong, J., Chen, D., Tour, J.M.: Single-atomic ruthenium catalytic sites on nitrogen-doped graphene for oxygen reduction reaction in acidic medium (2017). https://doi.org/10.1021/acsnano.7b02148
89. Wu, X., Mu, F., Lin, Z.: Three-dimensional printing of graphene-based materials and the application in energy storage. Mater. Today Adv. **11**, (2021). https://doi.org/10.1016/j.mtadv.2021.100157
90. Kurian, M.: Recent progress in the chemical reduction of graphene oxide by green reductants–a mini review. Carbon Trends. **5**, 100120 (2021). https://doi.org/10.1016/j.cartre.2021.100120

Opinions on Graphene as a Super-Versatile Material for the Near Future

Sachin Sharma Ashok Kumar, Shahid Bashir, Ramesh Kasi, and Ramesh T. Subramaniam

Abstract The discovery of graphene in year 2004 triggered the research undertakings at various theoretical and experimental aspects globally due to its outstanding properties and extremely large surface area. In the recent years, graphene and graphene-based materials have proven to be a rich source for the construction and utilization of it in many fields. Globally, significant efforts have been devoted by researchers to investigate the properties of graphene and extend its applications. Up to date, there are several research activities that are being conducted in relation to graphene and graphene-based materials in the application of various energy storage and conversion devices, sensors, photocatalysis, water purification, drug delivery etc. In our opinion, although, graphene has proven its versatility in every branch of science and technology, more research efforts are deemed necessary to harness the full potential of graphene and graphene-based materials for a wide range of applications. For instance, for future work, the application of graphene must be further extended to cover all the promising fields.

Keywords Graphene · Specific properties · Expert opinion · Challenges · Future perspectives

In the recent years, graphene and graphene-based materials have proven to be a rich source for the construction and utilization of bio-electrochemical and bioelectronics sensors. Illustrated in Table 1 are the countless examples of graphene based electrochemical and electrical based biosensors [1–17]. However, there are still various challenges that needs to be overcome despite the development and advancement of

S. S. Ashok Kumar · S. Bashir · R. Kasi (✉) · R. T. Subramaniam
Department of Physics, Faculty of Science, Centre for Ionics Universiti Malaya, 50603 Kuala Lumpur, Malaysia
e-mail: rameshkasi@um.edu.my

S. Bashir
Higher Institution Centre of Excellence (HICoE), UM Power Energy Dedicated Advanced Centre (UMPEDAC), Level 4, Wisma R&D, Universiti Malaya, Jalan Pantai Baharu, 59990 Kuala Lumpur, Malaysia

© The Author(s), under exclusive license to Springer Nature Singapore Pte Ltd. 2023
R. T. Subramaniam et al. (eds.), *Graphene*, Engineering Materials,
https://doi.org/10.1007/978-981-99-1206-3_9

graphene materials that plays a vital role in enhancing the electrical and electrochemical biosensor attributes. In addition, to bring some of the current sensors into the real biomedical applications, there is a huge necessity to move beyond research to develop new concepts which will result in a better selectivity and sensitivity respectively. Up to date, numerous researches have been conducted and reported on graphene-based sensors, whereby it has exhibited good sensing attributes, however, the novelty on its performance in real biological samples where the high salt concentrations are present next to a variety of other proteins are largely lacking [18]. Since it is known that the primary concern on graphene-based interfaces are the non-specific interactions of these proteins, therefore, this insufficient information is one of the vital limiting factors for current commercialization. Illustrated in Fig. 1 are the main research focus for graphene-based bioelectronics platforms in the recent years and the examples of the countless utilization of various types of graphene materials [18].

Globally, significant efforts have been devoted by researchers to integrate polyethylene modified pyrene ligands and block with the human serum itself, thus resulting in the enhancement of the sensor's selectivity particularly when utilized in complex biological media [18]. Additionally, the current limitation of the synthetization of large-scale reproducibility of graphene biosensor interfaces is another limiting factor for commercialization. On the other hand, in terms of bioelectronics, to enable for sensing in high ionic strength solutions namely saliva, serum, etc., there is a need to mitigate the overbearing sensitivity-limiting factor from the Debye screening length. In other words, an insight view of the aptamer or analyte affinity strength and interaction become a fundamental issue to ensure favourable bioelectronics detection of a variety of proteins [18]. Therefore, the change from the well-known and highly specific antibodies strategies to less selective aptamer technology is significantly important for future applications.

Interestingly, compared to existing conventional materials, graphene and its derivatives have shown their versatility in photovoltaic devices due to their unique properties. For device performance, a key to achieve practical applications is the synthesis-tailored structure to-properties of graphene-based materials. Furthermore, graphene as a conductive electrode is a promising substitute for commercial indium doped tin oxide leading to flexible solar cells. In solar cell buffer layers, it has been revealed that the graphene-based materials have a high capability to function as charge selective and transport components [19]. Although, there have been remarkable progresses in graphene-based photovoltaic devices, there are still many challenges that need to be addressed in the future. Firstly, the mass production of continuous large size graphene will require an advanced reliable growth approach. Secondly, another vital issue is to integrate the graphene-based materials into the functional devices. To control the properties during device fabrication, the fully integrated separation and purification processes are utilized. In addition, there is a need to develop a novel with a simple and accurate method to produce the derivatives of graphene with controlled properties even though the functionalization (chemically, thermally), size control, band energy tuning of graphene for a wide variety of applications have been achieved. With the new implementation, a new generation of devices result with fascinating features such as lightweight, eco-friendly and low cost

Table 1 Summary of the recent examples of bio-electrochemical and bioelectronics graphene-based sensors

Analyte	Electrode	Technique	Linear range	LoD	Remarks	References
H_2O_2	rGO-TiO_2	Chronoamperometry (CA)	0.1–360 μm	10 nm	hemoglobin	[17]
H_2O_2	rGO/Platinum nanoparticles	CA	0.2 nm–3 μm	20 nm	Chitosan-ferrocene Carboxylic acid	[16]
Prostate specific antigen (PSA)	NH_2-graphene	Electrochemical impedance spectroscopy (EIS)	2 pg/ml–2 μg/ml	0.46 pg/ml	anti-PSA and gold nanoparticle amplification	[11]
PSA	rGO	Differential pulse voltammetry (DPV)	90 ng/ml–0.1 pg/ml	10 fg/ml	aptamer	[10]
VEGF165	rGO	Single sweep potential (SWP)	10 fg/ml–1 ng/ml	8 fg/ml	Cucurbituril and N_3 modified rGO + aptameter + alkynyl-DNA	[13]
Folic acid	rGO/MoS_2	DPV	0.01–100 μm	10 nm	Direct oxidation	[15]
Nucleic acid	Thiofluoro-graphene	EIS			Nucleic acid adsorption	[8]
Carcinoembryonic antigen (CEA)	Graphene	Field effect transistors (FET)		0.5 pm	1-pyrenebutanoic acid succinimidyl ester (PASE) and anti-CEA	[5]
Folic acid protein (FAP)	rGO (Electrophoretic deposition, EPD)	DPV	1–100 pm	1 pm	Modified folic acid	[3]
Single nucleotide polymorphism (SNP)	Graphene	FET	100 nm–100 μm	100 nm	1-pyCOOH	[7]
Giladin	Porous rGO (prGO)	DPV	1.2–34 ng/ml	1.2 ng/ml	1-pyrenebutyric acid (PA) and anti-giladin	[1]

(continued)

Table 1 (continued)

Analyte	Electrode	Technique	Linear range	LoD	Remarks	References
Glucose	rGO/N-doped prGO/CUO	CA	0.25 μm–6 mm	0.25 μm	EPD	[4]
Dopamine	rGO/PEI (EPD)	DPV		50 nm	Microsystem	[14]
Thrombin	Laser-scribed graphene electrode	DPV	1 pm–0.1 nm	1 pm	PA and thrombin aptamer	[12]
Lysozyme	prGO	DPV	0.05–7.5 μm	50 nm	Direct oxidation	[2]
Micro-RNA	rGO	FET		10 Fm	Gold nanoparticles modified PNA and 1-pyCOOH	[6]
Interleukin-8	rGO	DPV	500 fg/ml–4 ng/ml	72.7 pg/ml	Gold nanoparticles and anti-IL8	[9]

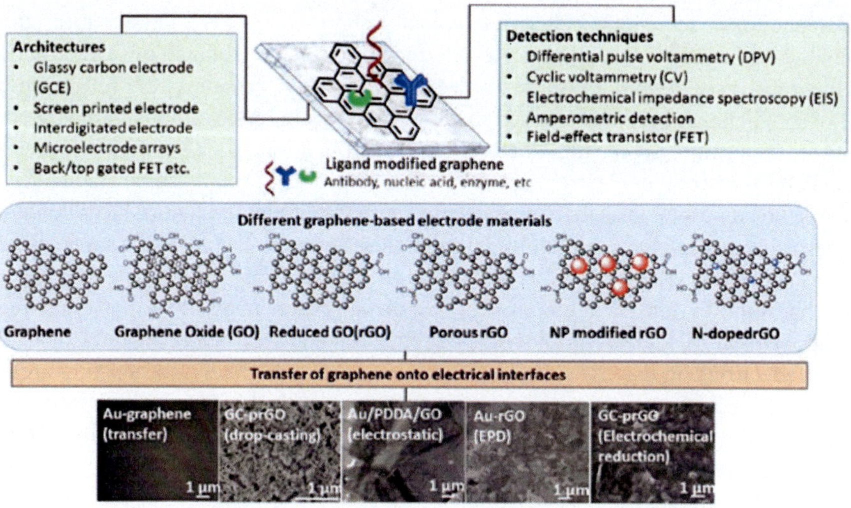

Fig. 1 Application of different forms of graphene for bio-electrochemical and bioelectronics sensors. Reproduced with permission Szunerits and Boukherroub [18]

respectively [19]. In short, due to its environmental stability, photo-catalytic properties, flexibility and high electrical conductivity, graphene has a lot more to offer in photovoltaic technology if engineered carefully to meet specific requirements.

Up to date, there are several research activities that are being conducted in relation to graphene and graphene-based materials in the application of various energy storage or conversion devices [19]. In addition, compared to other carbon allotropes, graphene and GO have been more significant candidates due to key merits such as two-dimensional networks and good hydrophobicity, thus resulting it to suitable to enhance the electrochemical performances [20]. Moreover, other two-dimensional materials such as germanene, different transition metal chalcogenides, hexagonal boron nitride, borophene, silicene, stanene, phosphorene and boron nitride nanosheets respectively have been inspired by graphene and its derivatives, hence, resulting in the emergence of these materials as new nanostructures for the future applications [19]. For instance, it has been reported that the applications of graphene and its derivatives in the field of energy storage has been fascinating. Here, graphene was observed to exhibit good electrochemical properties due to its extraordinary properties namely high surface area, high mechanical and chemical stability and controllable morphological features. Lately, graphene derivatives namely GO, doped graphene, 3D graphene, nano-porous graphene, nano-mesh graphene etc., have been reported to exhibit better electrochemical properties compared to pristine graphene. Furthermore, the hexagonal boron nitride was observed to have a low electrical conductivity, however it exhibited superior energy storage capacity. Therefore, in order to enhance their properties, graphene materials were incorporated, thus, resulting the electrochemical performance as the electrode, electrolyte, and separator in secondary batteries to enhance significantly [19]. However, in our opinion, although, graphene has proven its versatility in every branch of science and technology, more research efforts are deemed necessary to harness the full potential of graphene and graphene-based materials for a wide range of applications. For instance, for future work, the electrochemical application of GO must be further extended to cover other electrochemical applications. Furthermore, some challenges such as the facile synthesis that are associated with the GO material must be critically addressed and it is also essential to determine the structure of GO at the molecular level. In addition, the effect of the defects on the conductivity of GO is another key area that requires further attention [21]. On an important note, to further improve the application of GO in various other applications, a deep understanding of the flow of electrons on the GO substrate interface will also be a ground-breaking area of research. For the future advancement of this material, the design and approach adopted in the fabrication of GO-based device is also critical. In short, prior upon highlighting these challenges, the GO material remains a key future in the application of electrochemical sensors.

In addition, the costing of graphene and its derivatives is another important aspect that will be further discussed. For instance, supported on SiO_2 substrate, it has been previously reported that the price of the monolayer and bilayer graphene sheets that have been micromechanically produced which were in the range of £ 0.5 to £ 3 per μm^2 [22, 23]. Recently, it was revealed that the cost of graphene sheets produced in large quantities were significantly lower compared to carbon nanotubes (CNTs) [24–31]. Furthermore, for the research grade and commercial CNTs, the price for multi-walled and single-walled CNTs was estimated at USD 1 to USD 2 per gram and to hundreds (USD) per gram respectively. As mentioned in the previous chapters, in order to produce useful graphene-based materials, the availability of graphene sheets in large quantities is significantly essential. In year 2006, several researchers globally reported that the price of graphite nano-platelets to produce graphene sheets through various methods such as bottom-up, epitaxial growth, mechanical cleavage and chemical exfoliation of graphite respectively was estimated to be around USD 10 per kg [32].

Alternatively, the average cost of GO, a derivative of graphene, was estimated between USD 20 and USD 30 per kg, which was found to be higher than graphite nanoplatelets [32]. This is due to the need of GO to be exfoliated by either chemical or thermal treatments [33, 34]. Moreover, in the recent years, the thermally reduced GO are commercially sourced by two companies, Vorbeck and Angstron materials respectively. Compared to pure graphene, the GO has been reported to be more costly whereby the price range is between USD 300 and USD 800 per kg respectively. However, the cost reduction of graphene materials production can be achieved by recycling the acid wastes and to scale up the production of graphene by continuously fluidizing the bed reactor for thermally reduced GO [35]. In short, being atomically thin, the low cost and fascinating properties of graphene sheets make it a promising and a super-versatile material as a nanoscale building block for new material and robust components for nano-electronic and nano-electromechanical devices as well as a wide range of other applications for the near future.

In conclusion, the discovery of graphene in year 2004 triggered the research undertakings at various theoretical and experimental aspects globally due to its outstanding properties and extremely large surface area [20]. Prior to its discovery, owing to their resourceful applications in a wide range of industries, graphene and graphene-based materials have enjoyed the status of being the new super-versatile materials on the horizon of condensed matter physics and material science [36]. In the year 2020 and 2021, the global graphene market size was valued at approximately USD 285.9 million and USD 388.8 million, respectively, and it is anticipated that the compound average annual growth rate (CAGR) will rise up to 44.2% during the forecast period from 2021 to 2030 [37]. Therefore, in our opinion, due to their superior properties namely electrical and thermal conductivity, high electron mobility, strength, toughness, durability, lightweight, flexible, and high permeability, we predict that there will be a tremendous growth and expansion in graphene research and development in the coming years around the world for various applications.

Acknowledgements Authors would like to thank Universiti Malaya for providing the research facilities and IIRG grant IIRG007C-19IISS and SATU Joint grant ST031-2021.

Author Contributions Sachin Sharma Ashok Kumar wrote the original draft, Shahid Bashir edited the original draft, K. Ramesh supervised, reviewed, and edited the draft, and S. Ramesh supervised, reviewed, and edited the draft.

Conflict of Interest Authors have no competing interests.

References

1. Chekin, F., et al.: Reduced graphene oxide modified electrodes for sensitive sensing of gliadin in food samples. ACS Sens. **1**(12), 1462–1470 (2016)
2. Vasilescu, A., et al.: Porous reduced graphene oxide modified electrodes for the analysis of protein aggregation. Part 1: Lysozyme aggregation at pH 2 and 7.4. Electrochim. Acta **254**, 375–383 (2017)
3. He, L., et al.: Detection of folic acid protein in human serum using reduced graphene oxide electrodes modified by folic-acid. Biosens. Bioelectron. **75**, 389–395 (2016)
4. Maaoui, H., et al.: Copper oxide supported on three-dimensional ammonia-doped porous reduced graphene oxide prepared through electrophoretic deposition for non-enzymatic glucose sensing. Electrochim. Acta **224**, 346–354 (2017)
5. Zhou, L., et al.: Label-free graphene biosensor targeting cancer molecules based on non-covalent modification. Biosens. Bioelectron. **87**, 701–707 (2017)
6. Cai, B., et al.: Gold nanoparticles-decorated graphene field-effect transistor biosensor for femtomolar MicroRNA detection. Biosens. Bioelectron. **74**, 329–334 (2015)
7. Hwang, M.T., et al.: Highly specific SNP detection using 2D graphene electronics and DNA strand displacement. Proc. Natl. Acad. Sci. **113**(26), 7088–7093 (2016)
8. Urbanová, V., et al.: Thiofluorographene–hydrophilic graphene derivative with semiconducting and genosensing properties. Adv. Mater. **27**(14), 2305–2310 (2015)
9. Verma, S., et al.: Anti-IL8/AuNPs-rGO/ITO as an immunosensing platform for noninvasive electrochemical detection of oral cancer. ACS Appl. Mater. Interfaces **9**(33), 27462–27474 (2017)

10. Kavosi, B., et al.: Ultrasensitive electrochemical immunosensor for PSA biomarker detection in prostate cancer cells using gold nanoparticles/PAMAM dendrimer loaded with enzyme linked aptamer as integrated triple signal amplification strategy. Biosens. Bioelectron. **74**, 915–923 (2015)
11. Yan, M., et al.: A disposable electrochemical immunosensor based on carbon screen-printed electrodes for the detection of prostate specific antigen. Biosens. Bioelectron. **38**(1), 355–361 (2012)
12. Fenzl, C., et al.: Laser-scribed graphene electrodes for aptamer-based biosensing. ACS Sens. **2**(5), 616–620 (2017)
13. Wei, T., et al.: Cucurbituril and azide cofunctionalized graphene oxide for ultrasensitive electro-click biosensing. Anal. Chem. **89**(22), 12237–12243 (2017)
14. Kahlouche, K., et al.: Controlled modification of electrochemical microsystems with polyethylenimine/reduced graphene oxide using electrophoretic deposition: sensing of dopamine levels in meat samples. Talanta **178**, 432–440 (2018)
15. Chekin, F., et al.: MoS_2/reduced graphene oxide as active hybrid material for the electrochemical detection of folic acid in human serum. Biosens. Bioelectron. **85**, 807–813 (2016)
16. Bai, Z., et al.: Non-enzymatic electrochemical biosensor based on Pt NPs/RGO-CS-Fc nanohybrids for the detection of hydrogen peroxide in living cells. Biosens. Bioelectron. **82**, 185–194 (2016)
17. Liu, H., et al.: Hollow TiO_2 modified reduced graphene oxide microspheres encapsulating hemoglobin for a mediator-free biosensor. Biosens. Bioelectron. **87**, 473–479 (2017)
18. Szunerits, S., Boukherroub, R.: Graphene-based bioelectrochemistry and bioelectronics: a concept for the future? Curr. Opin. Electrochem. **12**, 141–147 (2018)
19. Ashok Kumar, S.S., et al.: A review on graphene and its derivatives as the forerunner of the two-dimensional material family for the future. J. Mater. Sci. 1–43 (2022)
20. Kumar, S.S.A., et al.: A comprehensive review: super hydrophobic graphene nanocomposite coatings for underwater and wet applications to enhance corrosion resistance. FlatChem 100326 (2021)
21. Zhang, B., et al.: Mechanical and thermal properties of hierarchical composites enhanced by pristine graphene and graphene oxide nanoinclusions. J. Appl. Polym. Sci. **131**(19), (2014)
22. Kumar, S.S.A., et al.: New perspectives on Graphene/Graphene oxide based polymer nanocomposites for corrosion applications: the relevance of the Graphene/Polymer barrier coatings. Prog. Org. Coat. **154**, 106215 (2021)
23. Tang, B., et al.: Application of graphene as filler to improve thermal transport property of epoxy resin for thermal interface materials. Int. J. Heat Mass Transf. **85**, 420–429 (2015)
24. Othman, N.H., et al.: Graphene-based polymer nanocomposites as barrier coatings for corrosion protection. Prog. Org. Coat. **135**, 82–99 (2019)
25. Stankovich, S., et al.: Synthesis of graphene-based nanosheets via chemical reduction of exfoliated graphite oxide. Carbon **45**(7), 1558–1565 (2007)
26. Stankovich, S., et al.: Stable aqueous dispersions of graphitic nanoplatelets via the reduction of exfoliated graphite oxide in the presence of poly (sodium 4-styrenesulfonate). J. Mater. Chem. **16**(2), 155–158 (2006)
27. Stankovich, S., et al.: Graphene-based composite materials. Nature **442**(7100), 282–286 (2006)
28. Niyogi, S., et al.: Solution properties of graphite and graphene. J. Am. Chem. Soc. **128**(24), 7720–7721 (2006)
29. Xu, Y., et al.: Flexible graphene films via the filtration of water-soluble noncovalent functionalized graphene sheets. J. Am. Chem. Soc. **130**(18), 5856–5857 (2008)
30. Li, D., et al.: Processable aqueous dispersions of graphene nanosheets. Nat. Nanotechnol. **3**(2), 101–105 (2008)
31. Eda, G., Fanchini, G., Chhowalla, M.: Large-area ultrathin films of reduced graphene oxide as a transparent and flexible electronic material. Nat. Nanotechnol. **3**(5), 270–274 (2008)
32. Fukushima, H., et al.: Thermal conductivity of exfoliated graphite nanocomposites. J. Therm. Anal. Calorim. **85**(1), 235–238 (2006)

33. Kumar, S.S., et al.: Introducing graphene thin films into carbon fiber composite structures for lightning strike protection. Polym. Compos. **40**(S1), E517–E525 (2019)
34. Ashok Kumar, S.S.: Incorporation of graphene thin films into the carbon fiber reinforced composite via 3d composite concept against the lightning strikes on composite aircraft. Wichita State University (2012)
35. Lee, S., et al.: Large-scale production of high-quality reduced graphene oxide. Chem. Eng. J. **233**, 297–304 (2013)
36. Sharma, S.S.A., et al.: The significance of graphene based composite hydrogels as smart materials: a review on the fabrication, properties, and its applications. FlatChem 100352 (2022)
37. Kumar, S.S.A., et al.: Why is graphene an extraordinary material? A review based on a decade of research. Front. Mater. Sci. **16**(2), 1–39 (2022)